# Introductory Microbiology

## ABOUT THE AUTHORS

**D. Balachandar** is working as Associate Professor in the Department of Agricultural Microbiology, Tamil Nadu Agricultural University, Coimbatore since 1997. His fields of specialization are microbial genetics, diversity of bacteria in different ecosystems, soil genomics and biological nitrogen fixation. He has over 10 years of research and teaching experience. He has published about 20 research articles in national and international journals and 5 book chapters. He was awarded with UNESCO and DBT overseas fellowships for training in Biotechnology in Japan and USA.

**R. Thamizh Vendan** was awarded Ph.D., degree in Agricultural Microbiology from Tamil Nadu Agricultural University, Coimbatore. Presently he is working as Associate Professor in the Agricultural College & Research Institute, (*A Constituent College of TNAU, Coimbatore)*, Tiruchirappalli, Tamil Nadu. His fields of specialization are soil microbes - plant interaction, biofertilizers and liquid inoculants. He has over 14 years experience of teaching and research. He was awarded with 'Young Agricultural Scientist' award in the year 2000. He had been to International Rice Research Institute (IRRI), Philippines for training and to Thailand for Biotechnology conference. More than 25 research articles by him have been published in national and international journals.

# Introductory Microbiology

Authors

D. BALACHANDAR
Associate Professor
Department of Agricultural Microbiology
Tamil Nadu Agricultural University
Coimbatore - 641 003

&

R. THAMIZH VENDAN
Associate Professor (Microbiology)
Agricultural College and Research Institute
*(A Constituent College of TNAU, Coimbatore)*
Tiruchirappalli, Tamil Nadu

New India Publishing Agency
Pitam Pura, New Delhi - 110 088

*Published by*
Sumit Pal Jain *for*
New India Publishing Agency
101, Vikas Surya Plaza, CU Block, L.S.C. Mkt.,
Pitam Pura, New Delhi- 110 088, (India)
Phone: 011-27341717, Fax: 011-27341616
E-mail: newindiapublishingagency@gmail.com
Web: www.bookfactoryindia.com



ISBN (13) : 978-81-89422-78-3

Composed and Designed by NIPA

**Tamil Nadu Agricultural University**

**Prof. C.RAMASAMY**
Vice-Chancellor

COIMBATORE-641 003
TAMIL NADU
INDIA

# FOREWORD

The overarching supremacy of micro organisms in the biosphere is so evident despite their size, they are certainly of immense importance to man and most of them are indispensable to his survival on earth Microbes have been widely used as subjects for the study of molecular and cellular biology but they are also interesting biologic forms in their own right and have been best known through the decades for their helpful and harmful activities. Microbiology is a subject that impinges on almost every aspect of human existence and the study of microbiology at undergraduate level has become necessary in the University syllabi.

The authors of the textbook **"Introductory Microbiology"**, Dr. D. Balachandar and Dr. R. Thamizh Vendan have made an attempt to arrange diverse areas of microbiology, in a concise but comprehensive manner. They have given sufficient and detailed information to inspire the students to progress to more advanced and specialized texts. I feel that a text book of this type is really appropriate for undergrduate students as it covers University syllabus. I appreciate the authors for bringing out this useful book.

Coimbatore
July 20, 2007

**(C. RAMASAMY)**

# Preface

At the outset, we were very daunted at the prospect of writing a small book to cover such a vast topic as microbiology. To attempt to cover the whole of the subject would have been an impossible task. Hence, this book is an introduction to general microbiology with a lucid text in simple language. In view of wide range of microbiology textbooks currently on offer from bookshops, one might reasonably question the need for another. The simple answer is that there is always a need for introducing new texts for updating the students to the current scenario. This book has been presented in sixteen chapters and they have been selected based on the University syllabus of General Microbiology. It has been essentially written to cater the needs of both undergraduate and postgraduate students of Indian Universities. We hope that this book may be useful as a primer for students embarking on more specialized areas in microbiology.

We owe an intellectual debt to numerous authors, editors and publishers whose information provided us the base material for the preparation of this text. We are grateful to Dr. C. Ramasamy, Vice Chancellor, Tamil Nadu Agricultural University, Coimbatore, who have been kind enough to write the foreword to this book. We thank Dr. M. Thangaraju, Director (Students Welfare), TNAU, Coimbatore and Dr. S. Jebaraj, Dean, Agricultural College & Research Institute, Tiruchirappalli for their support and encouragement.

The publisher of this book deserves a great deal of gratitude from us, who has given a nice shape to our text and expedited the publication. We have tried to keep errors in the text to the minimum and we shall welcome constructive suggestions for further improvement.

July 25, 2007
Coimbatore

**D. BALACHANDAR**
**R. THAMIZH VENDAN**

# CONTENTS

CHAPTER 1

# Introduction

## Microbiology

Study of microorganisms, a large and diverse group of microscopic organisms, that exist as single cell or cluster. It also includes virus, which are microscopic, but not cellular.

| Microbial cell | Plant & animal cell |
|---|---|
| Can live alone | Can exist only as part of organelle |
| Growth, energy generation and reproduction are independent | Depends other cells |

One of the first things to realize about the microbial world is that microorganisms are everywhere. Almost every natural surface is colonized by microbes (including your skin). Some microorganisms can live quite happily in boiling hot springs, whereas others form complex microbial communities in frozen sea ice.

Most microorganisms are harmless to humans. You swallow millions of microbes every day with no ill effects. In fact, we are dependent on microbes to help us digest our food and to protect our bodies from pathogens. Microbes also keep the biosphere running by carrying out essential

1. *Bacteria* are phylogenetically related group of unicellar prokaryotic organisms distinct from archeae.
2. *Archaea* are phylogenetically related group of prokaryotes which are primitive and distinct from bacteria.
3. *Fungi* are group of eukaryotic organisms lack of chlorophyll. They range in size and shape from single celled yeast to multicellular mushrooms.
4. *Algae* refers the group of eukaryotic organisms with chlorophyll. They range in size and shape from single celled algae (Ex: *Chlorella*) to complex cellular structured plant like algae (Ex. Kelp).
5. *Protozoa* are group of eukaryotic organisms lack of cell wall. The morphology, nutrition and physiology are different from other groups.
6. *Viruses* are group of non-cellular organisms, parasite or pathogen to plant, animals and other micro-organisms. They are too small and can be visualized only under electron microscopes.

## Need for Microbiology

- The microorganisms are research tool to understand the chemical and physical basis of life.
- They are tool to analyze the biochemical and genetic background of living things.
- They are excellent model for understanding the cell functions.
- They play important role in the field of medicine, agriculture and industry.

## Branches of Microbiology

*Medical Microbiology:* deals with the microorganisms associated with human body; diseases; diagnostic procedures; preventive measures.

*Aquatic microbiology:* deals with water purification; microbiological examination; biological degradation of wastes and ecology

*Aeromicrobiology:* deals with the microbes in air; contaminations; spoilage; dissemination of diseases

*Food microbiology:* deals with the microorganisms associated with foods; food borne pathogens; food spoilage; food preservation and food safety

*Agricultural Microbiology:* deals with plant associated microbes; soil fertility; plant and animal diseases; microbial degradation of organic matter; soil nutrients transformation

*Industrial microbiology:* deals with production of antibiotics and vaccines using microorganisms; fermentation and fermented products such as beverages, bread, milk products; production of protein and hormones by genetically modified organisms.

*Exomicrobiology:* Exploitation for life in outer space

*Geochemical microbiology:* deals with coal, gas and mineral formation; mineral recovery through microbes from ore (bioleaching).

CHAPTER

2

# History of Microbiology

Obviously human have had to deal with microbes even before the recorded history. The first record of human using microorganisms comes from ancient tablets from mid east.

Babylonians were using yeast to make beer over 8000 years ago and acetic acid bacteria to make vinegar over 6000 years ago. About 5000 years ago, Persia (Now Iran) region recorded the wine making.

The Romans had God for specific microorganisms. The roman God of mold and mildew were "*Robigus*" and "*Robigo*" which mean crop rust. (Rust is one of the plant disease caused by fungus). God Robigus was very much feared because of crop lost. About 2000 years ago, Romans proposed that diseases were caused by tiny animals. But, fundamentalist religions had a strong hold over the progress. The real microbiology history starts from 1600s, when people began to make crude lenses and microscopes.

The following are the scientists and their role in the development of microbiology:

## Discovery of Microscope

*Robert Hooke (1635 - 1700)* : One maker of microscope (30x magnification) discovered the cell (Latin: cellulae meaning small room) from cross sections of cark. He noticed some microscopic fungi too.

Robert Hooke's Microscope

*Antony Van Leeuwenhock (1632 - 1723)* : Lived during same age of Robert Hooke, was an amateur lens and microscope maker. His microscope can magnify 300x times. He was the first to describe the protozoa and bacteria. He observed some bacteria from plagues of his own teeth. He named them as *animacules*. He was way ahead of his time and many scientists too did not believe him. But now, he recognized as Father of Microbiology.

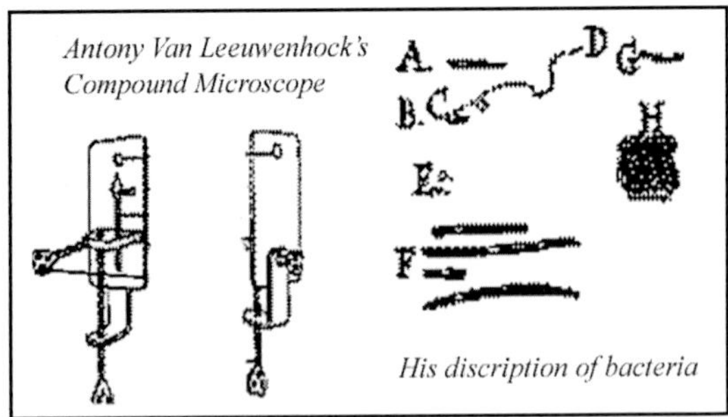


Van Leeuwenhoek's microscope and his description of bacteria

## Disprove of Spontaneous Generation Theory

At that time, the age old idea of "Spontaneous Generation theory" was the dominant one. The idea that organism originate directly from non-living matter. (Life from non-living) also called as *abiogenesis* (a – not; bio – life; genesis – origin).

*Ex :* Maggots were developed spontaneously via recombination of matters in rotting materials. (ex meat)

*Francesco Redi (1626 – 1697) :* Conducted experiment to disprove the SG theory. He placed a meat in a jar and covered with net and sealed. The flies laid the egg on the cover/net and maggots developed on the cover. So He established that the origin of maggot was from fly only not from meat.

*Lazaro Spallanzani (1729 – 1799) :* boiled the beef broth for an hour and sealed in a flask. No microbes observed during incubation.

*(But till the period, the supporters of SG theory believed that air is essential for SG).*

*Franz Schulze (1815 – 1873) :* passed the air through strong acid before contact to beef broth leads no microbial growth during incubation.

*Theodor Schwann (1810 – 1882) :* passed the air through red hot tube before contact to beef broth leads no microbial growth during incubation.

*(But now they thought that heat and acid alter the air leads no SG in beef broth).*

*H. Schroder & T. Von Dusch (1850) :* used cotton plug to filter the air on the mouth of the flask and observed no growth. (Cotton plug for microbial culture and media preparation technique was initiated from now).

*Louis Pasteur (1822 – 1895)* : performed "gooseneck experiment". The nutrient of flask was heated and the untreated - unfiltered air could pass in or out, but the germs settled in the gooseneck and no microbes were observed in the nutrient solution.

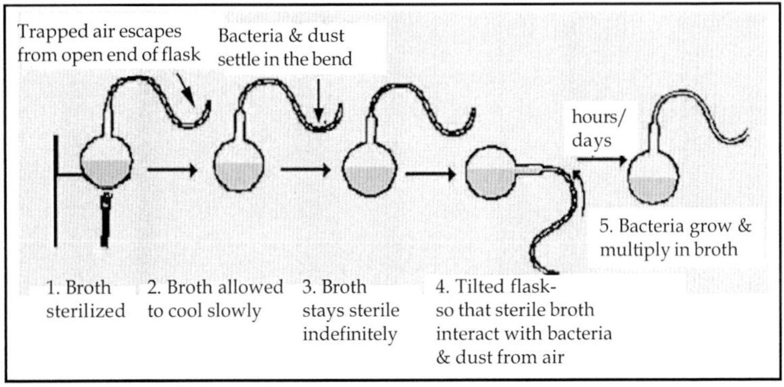


Gooseneck experiment of Louis Pasteur

His concept of Germs theory of disease (means germs are responsible for the disease not the inert mater) ends the SG theory.

*John Tyndall (1820 -1893)* : proved that dust carries the germs and if no dust in the air, the sterile broth remained free of microbial growth for indefinite period.

He also developed a sterilization method "Tyndallization", referred as intermittent or fractional sterilization. The subsequent cooling and heating by steam for 3 days will remove the germs and their spores.

*Contributions of Louis Pasteur (1822 – 1895)* : Chemistry Professor, France

- *Disproved* the SG theory.
- Discovered that *fermenting* fruit to alcohol by microbes - From that Fermentation started.
- *Sorted* different microbes giving different taste of wine.

- He selected a particular strain (*Yeast*) for high quality wine.
- He developed a method to remove the undesired microbes from juice without affecting its quality . Heating the juice at 62.8°C for half-an hour did the job. This technique is called as *Pasteurization,* which is commonly used in the field of milk industry.
- He discovered that parasites (protozoa) causing *pebrine* disease of silk worm. He suggested that disease free caterpillars can eliminate the disease.
- He isolated the *anthrax* causing bacilli from the bloods of cattle, sheep and human being.
- He also demonstrated the *virulence* (ability of microbe to cause disease) of bacteria.
- He developed *vaccine* (a killed or attenuated microbe to induce the immunity) against rabies from the brains and spinal cord of rabbit.

*Edward Jenner (1898) :* Developed vaccines for small pox disease. He coined the term vaccines.

*Robert Koch (1843 – 1810) :* Isolated bacilli from bloods of cattle and infected with healthy one and observed the disease. He proved that microorganisms causes the disease. He formulated a sequence procedure to prove that specific microorganism causes the specific disease, called as *Koch's Postulate.*

**Koch's postulates**

1. The agent must be present in every case of the disease.
2. The agent must be isolated and cultured *in vitro.*
3. The disease must be reproduced when a pure culture of the agent is inoculated into a susceptible host.
4. The agent must be recoverable from the experimentally-infected host.

He isolated tubercle bacillus (*Mycobacterium tuberculosis*). He developed a special staining to view *Mycobacterium* called *acid-fast staining*. He also discovered and isolated the cholera –causative agent, *Vibrio cholerae.*

*Walther Hesse & his wife Fannie E. Hesse (1883) :* used agar instead of gelatin for preparation of media. Agar goes to solution at 100°C and solidifies at 45°C. Till now this was not replaced by any other substance.

*Hans Chrisian Gram (1853 – 1933) :* developed a staining procedure for differentiating two different group of bacteria based on the cell wall structure (Gram +ve and Gram –ve ). The staining is called *Gram staining*.

*Edvin Klebs & Frederich Loeffler :* discovered the diphtheria bacilli and demonstrated the toxin production in the culture flask.

*Emil Von Behring & S. Kitasato :* devised to produce immunity by injecting the poison (toxin) into animals so that an antitoxin will be developed.

*Elic Metchnikoff :* demonstrated that the white blood cells can eat the disease causing bacteria in blood. He called the cells as *phagocytes* (eating cells) and the phenomenon as *phagocytosis*.

*Paul Erhlich :* explained the immunity of human. He also the first used a chemical (arsenic) to control the microbial growth (Chemotherapeutic agent)

*Joseph Lister (1878) :* developed *Pure culture* technique. Pure culture referred as the growth of moss of cells of same species in a vessel. He developed the pure cultures of bacteria using serial dilution technique.

He also discovered that *carbolic acid* to disinfect the surgical equipments and dressings leads the reduction of post-operational deaths/infections.

*Alexander Fleming (1928)* : identified *Penicillium notatum* inhibiting *Staphylococcus aureus* and identified the antibiotic *Penicillin.*

*Domagh, (1935)* : used *sulfanilamide* for chemotherapeutic treatment.

*Selman A Waksman, (1945)* : identified *Streptomycin* antibiotic from soil bacterium. He also coined the term antibiotics (referring a chemical substance of microbial origin which is in small quantity exert antimicrobial activity).

*Rene Dubos (1939)* : isolated *gramicidin* and *tyrocidin* from *Bacillus brevis.*

## Virus Research

- Unknowingly, *Edwart Jenner* handled the small pox virus and controlled by vaccines.
- Same time, *L. Pasteur* also handled the rabies virus. He also noticed that these microbes cannot be grown in lab condition and can not be seen under microscopes.
- *Charles Chambeland* developed the porcelain filter, which filter the bacteria but allow the rabies causative agent.
- In 1892, *Iwanowsk*i discovered the causative agent of tobacco mosaic disease, which was filterable. By spraying the bacteria free filtrate to healthy plant led to tobacco mosaic disease.
- 1935, *Wendell Stanley* crystallized the TMV. These crystals can cause the disease.
- 1936, *F.C.Bawden* explained that the crystalline powder contains protein and nucleic acid.
- *A.Gierer and G.Schramm* (1956) proved that the nucleic acid is responsible for infections.

*F.W.Twort and F.d'Herelle (1915) :* discovered the bacteriophages (bacterial viruses).

## Soil Microbiology

*Helriegel & Wilfarth :* found the symbiotic relation between legumes and bacteria.

*Leishman (1858) :* demonstrated that the nodules in legume were formed by bacteria.

*Muntz (1878) :* found the role of bacteria in nitrification in sewage water system.

*Martinus William Beijerinck (1851 – 1931)*

- Developed the enrichment technique to isolate various group of bacteria.
- Isolated sulphur reducing bacteria and sulphur oxidizing bacteria from soil.
- Isolated free-living nitrogen fixing bacterium, *Azotobacter* from soil.
- Root nodulating bacterium, Rhizobium, Lactobacillus, green algae were identified by him.
- He confirmed the Tobacco mosaic virus causes disease and it incorporated in the host plant to reproduce.

*Sergei Winogradsky (1856 – 1953)*

The following are the contributions of Winogradsky to soil microbiology.

- Microorganisms involved in N cycle, C cycle, S cycle.
- Nitrification process in soil.
- Autotrophy nutrition of bacteria.
- Chemolithotrophy nutrition of soil bacteria.
- Anaerobic nitrogen fixing *Clostridium pasteurianum.*

*J.G. Lipman and P.E.Brown (1903)* : explained the ammonification process in soil.

*L. Hiltner (1904)* : coined the term "*rhizosphere*" (refers the region around the root surface) where the microbial activity is very high.

*J. Dobereiner (1976)* : isolated *Azospirillum*, an associative nitrogen fixing organism which is commonly present in the grasses. The associative symbiotic Nitrogen fixation was started from here. She also isolated many associative nitrogen fixers from cereal roots.

*B. Frank (1885)* : coined the term *mycorrhiza* (association between fungus and higher plant roots).

*Barbara Mosse and J.W. Gerdemann* : identified the endomycorrhiza which can able to colonize the agricultural crops.

CHAPTER 3

# Introduction to Microscope

Microscope is an instrument to magnify and enable to see the microorganisms by us.

Microscopes are of two types. *Light* and *Electron* microscope.

## Basic Principles of light microscope

The light is the primary source on which magnification is performed. The magnification is obtained by a system of optical lenses using light waves. These light microscopes can be further classified into

a. Bright field microscope
b. Dark-field microscope
c. Fluorescence microscope and
d. Phase contrast microscope.

Basic units for microscope

1 meter = 1000 millimeter
1 millimeter = 1000 micrometer (mm) = $10^{-6}$ meter
1 micrometer = 1000 nanometer (nm) = $10^{-9}$ meter
1 Angstrom (1 A) = $10^{-8}$ meter
1 nanometer = 10 Angstrom

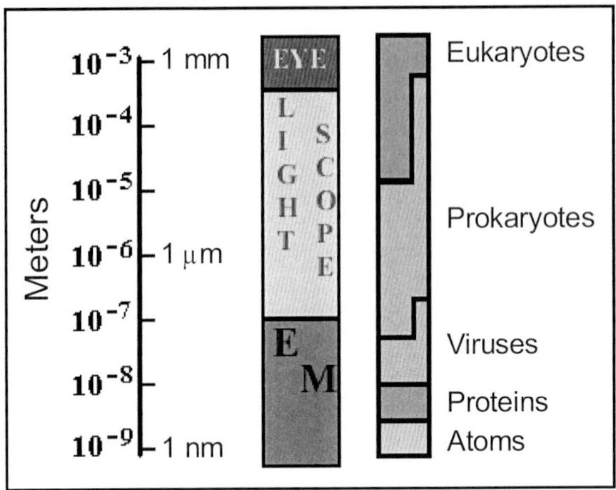


*Relative size of the micro organisms and their visibility.* Man can see about 0.5 mm sized object whereas the light microscopes can be used to visualize upto 1 mm and EM (electron microscopes) can be used to view 1 nm objects.

*Resolving Power :* The ability to distinguish two objects close to each other. The ability of a microscope in terms of resolving power is called as resolution.

*Ex :* Man has the resolving power of 0.2 mm (*meaning that he can distinguish two objects with a distance of 0.2 mm close to each other*) If he wants to see beyond the limit of his resolving power, further magnification is necessary. The magnification can be made by lenses.

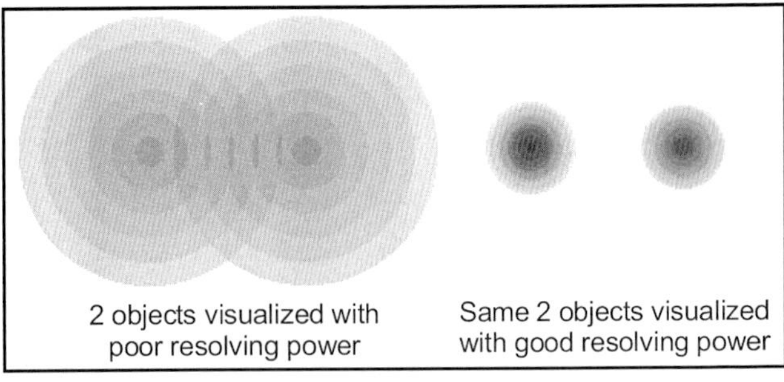


A good and poor resolving power

$$\text{Resolving power} = \frac{\lambda}{n\,(\sin\theta)}$$

where, λ is the wave length of light source and n (sin θ) [NA] is the *numerical aperture*

*(The resolving power (RP) should be of SMALLER value for good quality microscope)*

The resolving power of an microscope can be improved either by *reducing the wave length* of light or by *increasing the n (sin θ)* value.

*a. Reducing the λ*

This picture shows the range of spectrum of a light.

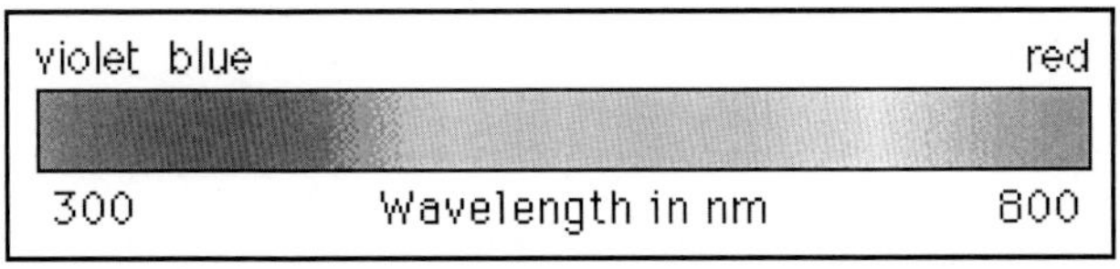

violet blue red

300 Wavelength in nm 800

The following are the wavelength of some colours: Blue - 400 nm; Red - 700 nm; Green - 550 nm

If we use violet light, we can get high resolution (meaning lowest resolving power), but, blue light (approx 450 nm) is sensitive to human being, the blue light is used in microscopes. The other lights will be removed from the light source using filters.

*b. Increasing NA (n sinθ)*

*Numerical aperture* measures how much light cone spreads out between condenser and specimen.

More spread of light gives less resolving power means better resolution.

The numerical aperture depends on the objective lens of the microscope. (objective lens refers the lens nearest to the specimen). There are two types of objective lenses available in any compound microscope.

a. *Dry objective lens :* which can view the specimen without any fluid. Air is the medium between the objective lens and the specimen.

b. *Oil-immersion objective lens :* which can view the specimen in the presence of immersion oil. This immersion oil is the medium between lens and the specimen.

*The oil immersion objective lens has more NA than dry objective lens.* Because, oil can increase the n value. In dry objective lens, the refractive index (n) is 1.0, because air is the medium, whereas, immersion oil has the refractive index (n) of 1.2 to 1.4. So, the resolving power of oil immersion objective lens is greater than dry objectives.

Ø = angle of light cone; maximum sin value is 1.0

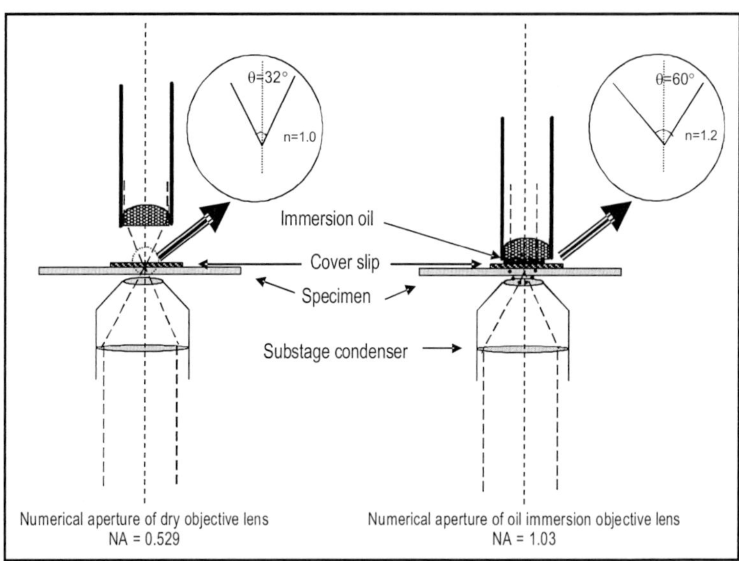


Reason for increased numerical aperture while using oil immersion objective lens

n = refractive index. n = 1.0 in air. It can be increased with certain oils (up to 1.4), called *immersion oil. (N.A. is property of lens. Look on side of lens to identify)*.

The sub-stage condenser (located below the specimen stage of the microscope) increased the angle of light cone too.

Theoretical limit of R.P. for light scope is *0.2 micrometers.*

*The limit of resolution:* The limit of resolution refers the smallest distance by which two objects can be separated and still be distinguishable or visible as two separate objects.

$$\text{The resolving power of compound microscope} = \frac{\lambda}{2NA}$$

## Calculation

*Resolving power*

For example If we use a microscope with green light (wavelength of 550 nm) and 0.2 NA objective lens in dry condition, the resolving power can be calculated as

$$RP = \frac{0.55}{2 \times 0.2} = 1.375 \text{ mm}$$

(Meaning that in this position, the two objects with a distance of 1.375 mm can be distinguishable).

*Magnification*

The magnification refers the number of times a specimen appeared to be larger than its original size.

For compound microscopes, the magnification can be calculated as follows:

$$\text{Magnification} = \text{objective lens magnification} \times \text{eye piece magnification}$$

If we use, 40x objective lens and 10x eye piece for viewing a specimen, then the magnification is 400 x. (refers that the specimen is magnified by 400 times than its original size).

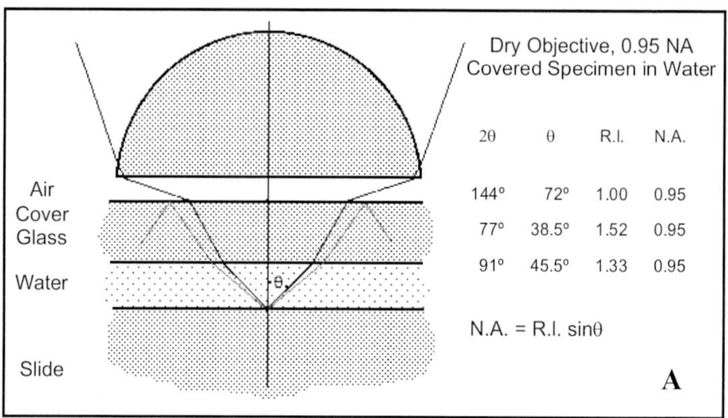


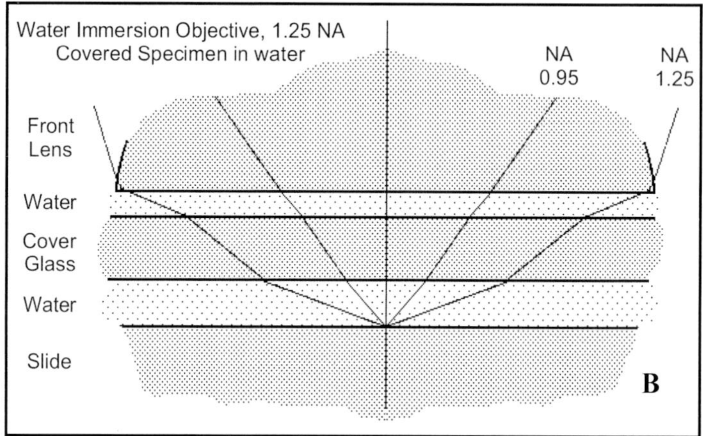


Numerical aperture of a dry objective lens when viewing a watery specimen with coverslip. (A) and with water immersion. (B) (if we use oil, the NA will be increased further).

## The Components of the light Microscope

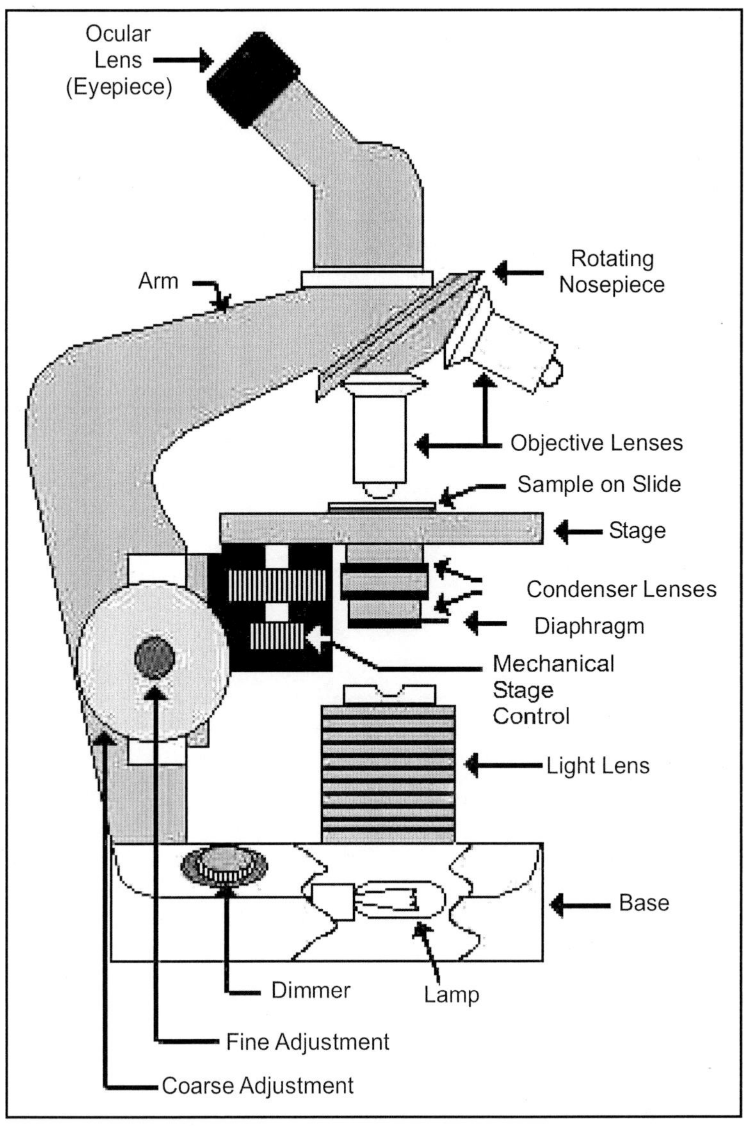


The Light Microscope: Optical Diagram.

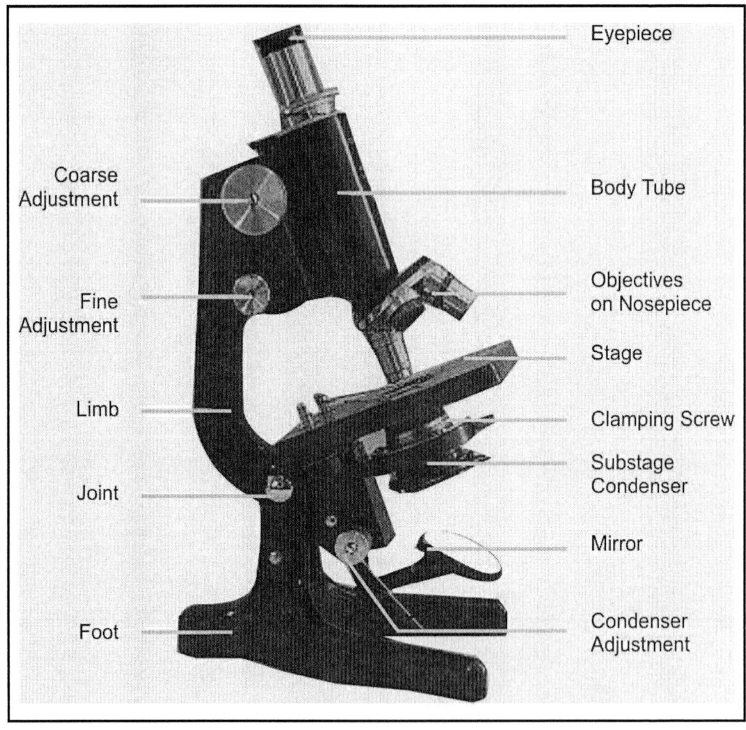


Parts of typical compound microscope

The optical components are those of all compound microscopes, and even though this model (right side) has no integral lamp, this discussion will begin with the lamp, and proceed in sequence via the mirror to the eye.

## The Microscope Lamp

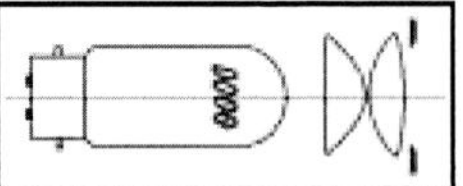

Microscopes require, especially at the highest powers, intense illumination. The intensity of a light source depends not so much upon its absolute power as upon the amount of light emitted from a given area of the source – lumens per square millimetre rather than just lumens. To achieve high intensity, various lamp designs have come and gone, but one of the commonest and most satisfactory is the low-voltage tungsten filament lamp, with the filament in the form of a tightly-wound flattened grid. The combination of a suitable quartz-halogen bulb with a concave spherical reflector works well and is also in wide use.

## The Mirror

The mirror is used only to fold the optical path of the microscope into a convenient space. It also introduces another source of potential mal adjustment into the system, and another surface to collect dust. Having said this, the mirror does not need to be of the highest optical quality to do its job, nor does a small amount of dust on the mirror make much difference to the quality of the image. A slight film of fine dust (such as remains after dusting the mirror with a blower brush) can actually be useful in locating the beam of light from the lamp when setting up the instrument. Always use the flat side of the mirror in combination with a substage condenser.

## The Sub stage Condenser

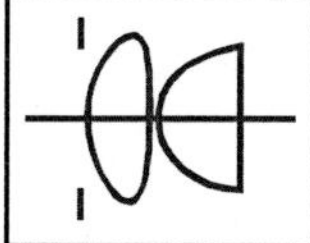

The sub stage condenser fitted to most microscopes is of a design originated by Ernst Abbe in the late 1800s and is usually referred to as the Abbe condenser. Whilst condensers of higher correction are available, the Abbe condenser has proven to be quite satisfactory for routine microscopy. A sub stage condenser of some kind is an absolute requirement for serious – or at least satisfactory – microscopy. The objectives from x20 upwards require the subject to be illuminated evenly over quite a large angle, and neither a concave mirror nor (especially) a flat mirror is capable of achieving this. If no condenser is used with a high power objective, the result is an image which is dark, coarse, contrasty and lacking in detail – described by earlier microscopists as "a rotten image". An important point to note here is that the sub stage condenser diaphragm is used to control the solid angle of the light emerging from the condenser, illuminating the specimen, and filling the objective – not for adjusting the brightness of the image.

## The Specimen

The specimen is usually supported by a slide and, essentially at the higher powers of the biological microscope, covered by a cover glass. Thus introduced into the image-forming light path, slide and cover glass become part of the optical system. The thickness of the slide is important to the correction of the sub stage condenser, and the thickness of the cover glass is critical to the performance of the objective, especially those of higher power (x20 and greater). Most sub stage condensers are corrected to work with a slide thickness of 1.0mm, and most microscope objectives of x20 or greater power are designed to work with a cover glass thickness of 0.17mm (thickness no. 1½).

## The Objective

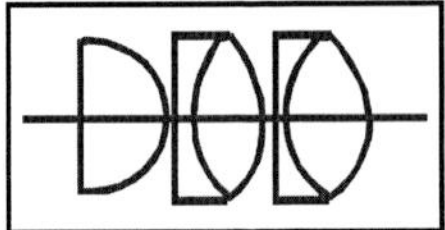

The objective is the most important component of the optical system in terms of the quality of the final image. The modern microscope objective probably represents the highest degree of optical perfection and precision engineering which is manufactured in volume for public consumption. The diagram shows a construction (not to scale) typical of a x40 achromatic objective standard on most laboratory microscopes.

The screw thread of microscope objectives has been a standard across the industry since 1858, when it was first proposed by the Royal Microscopical Society.

## The Eyepiece

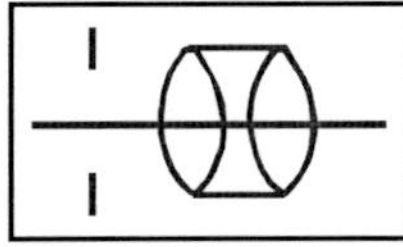

The eyepiece relays to the eye an image projected by the objective into the plane of the eyepiece diaphragm, further magnifying it in the process. In older microscopes, the eyepiece also corrected residual colour errors remaining in the objective. Modern infinity-tube length objectives are fully corrected in themselves, but still require additional focusing optics and appropriate eyepieces to produce their image. All of the apochromatic objectives of earlier period, and many of the higher power achromats, required compensating eyepieces. The other variety of eyepiece in common use is the Huyghenian – best suited to achromatic objectives in general, and particularly to low power achromats which often require little or no correction. These eyepieces are distinguished by a blue fringe around their field diaphragm when the eyepiece is used in bright field.

## The Eye

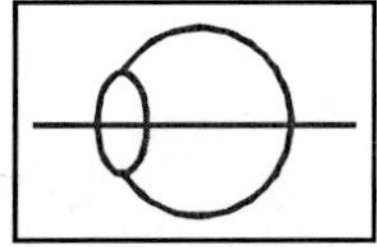

The cornea and the eye lens are the final optical components in the image-forming path to the retina. In a person with normal vision, the eye lens will be relaxed as though the eye is forming an image of a very distant object, and the focusing controls on the microscope used to achieve image sharpness. The optics of the eyepiece are such that all image-forming rays pass through a circle (called the Ramsden disc) a few millimetres exterior to the eyepiece lens and just smaller than the diameter of the pupil. The eye is bought close enough to the eyepiece for the ramsden disc and the pupil to coincide, at which point the full circular field of the microscope is seen. Learning to hold the head still in this optimum position, especially with a binocular instrument, is one of many skills acquired by the microscopist.

*Advantages:* convenient, relatively inexpensive, widely available.

*Disadvantages:* resolving power 0.2 micrometers at best, can recognize cells but not fine details ; Needs contrast; cells are mainly watery and don't contrast with their medium. Easiest way to view cells is to *fix* and *stain.*

| Optical Instrument | Resolving Power | RP in Angstroms |
|---|---|---|
| Human eye | 0.2 millimeters (mm) | 2,000,000 A |
| Light microscope | 0.20 micrometers (μm) | 2000 A |
| Scanning electron microscope (SEM) | 5-10 nanometers (nm) | 50-100 A |
| Transmission electron microscope (TEM) | 0.5 nanometers (nm) | 5 A |

## Types of Light Microscopy

The bright field microscope is best known to students and is most likely to be found in a classroom. Visible light is focused through a specimen by a condenser lens, and then is passed through two more lenses placed at both ends of a light-tight tube. The latter two lenses each magnify the image. Limitations to what can be seen in bright field microscopy are not so much related to magnification as they are to resolution, illumination, and contrast. Resolution can be improved using oil immersion lenses, and lighting and contrast can be dramatically improved using modifications such as dark field, phase contrast, and differential interference contrast. Fluorescence and confocal microscopes are specialized instruments, used for research, clinical, and industrial applications.

Other than the compound microscope, a simpler instrument for low magnification use may also be found in the laboratory. This is the stereo microscope, or dissecting microscope. Stereo microscopes usually have a binocular eyepiece tube, a long working distance, and a range of magnifications, typically from 5x to 35 or 40x.

## Bright Field Microscopy

With a conventional bright field microscope, light from an incandescent source is aimed toward a lens beneath the stage called the condenser, through the specimen, through an objective lens, and to the eye through a second magnifying lens, the ocular or eyepiece. Some microscopes have a built-in illuminator, while others use a mirror to reflect light from an external source. The condenser is used to focus light on the specimen through an opening in the stage. After passing through the specimen, the light is displayed to the eye with an apparent field that is much larger than the area illuminated. The magnification of the image is simply the objective lens magnification (usually stamped on the lens body) times the ocular magnification.

Students are usually aware of the use of the coarse and fine focus knobs, used to sharpen the image of the specimen. They are frequently unaware of adjustments to the condenser that can affect resolution and contrast. Some condensers are fixed in position, others are focusable, so that the quality of light can be adjusted. Usually the best position for a focusable condenser is as close to the stage as possible. The bright field condenser usually contains an aperture diaphragm, a device that controls the diameter of the light beam coming up through the condenser, so that when the diaphragm is stopped down (nearly closed) the light comes straight up through the center of the condenser lens and contrast is high. When the diaphragm is wide open the image is brighter and contrast is low.

## Microscopy with Oil Immersion

*Principle*

When light passes from a material of one refractive index to material of another, as from glass to air or from air to glass, it bends. Light of different wavelengths bends at different angles, so that as objects are magnified the images become less and less distinct. This loss of resolution becomes very apparent at magnifications of above 400x or so.

Placing a drop of oil with the same refractive index as glass between the cover slip and objective lens eliminates two refractive surfaces and considerably enhances resolution, so that magnifications of 1000x or greater can be achieved.

To use an oil immersion lens, first focus on the area of specimen to be observed with the high dry (400x) lens. Place a drop of immersion oil on the cover slip over that area, and very carefully swing the oil immersion lens and place over the oil drop. Focus carefully, preferably by observing the lens itself while bringing it as close to the cover slip as possible, then focusing by moving the lens away from the specimen. When in focus the lens nearly touches the cover slip. The focal plane is so narrow that it is very easy to focus right past it. If you are focusing toward the specimen, you can drive the lens right into it.

*When to use oil immersion lenses*

Use an oil immersion lens when you have a fixed (dead - not moving) specimen that is no thicker than a few micrometers. Even then, use it only when the structures you wish to view are quite small - one or two micrometers in dimension. Oil immersion is essential for viewing individual bacteria. It is nearly impossible to view living, motile cells at a magnification of 1000x, except for the very smallest and slowest.

A disadvantage of oil immersion viewing is that the oil must stay in contact, and oil is viscous. A wet mount must be very secure to use oil. Oil immersion lenses are used only with oil, and oil can't be used with dry lenses, such as 400x lens. Lenses of high magnification must be brought very close to the specimen to focus and the focal plane is very shallow, so focusing can be difficult. Oil distorts images seen with dry lenses, so once you place oil on a slide it must be cleaned off thoroughly before using the high dry lens again. Oil on non-oil lenses will distort viewing and possibly damage the coatings.

## Phase-Contrast and Dark Field Microscopy

A main obstacle in the microscopy of biological objects is their poor contrast. Only where a contrast exists or where it can be achieved by contrast-enhancing dyes, structures can be made visible. Light-absorbing parts of a preparation weaken the amplitude of the light waves that pass through them. It is thus also spoken of *amplitude preparations*. The change of stronger and feebler light is perceived by the eye as a difference in brightness. The invisible parts of the preparation are went through by the light without a change of amplitude, but the *phase* of the light may be changed depending on the consistency of the material. This change is due to the altered speed of the lightwaves. Differences in phase can be perceived neither by the eye nor by a photographical film.

The Dutch physicist F. Zernike succeeded in 1935 to convert phase to amplitude differences. He was awarded the Nobel price for this achievement in 1953. Today his method is known as phase-contrast microscopy and it is by now an integral part of nearly all research and many teaching microscopes. Its eminent advantage is that it allows the examination of living objects and thus to follow the processes within cells. It was the phase-contrast microscope that made it possible to make mitosis visible and even to film it.

For the procedure itself a special condenser with a ring-shaped mask and an additional "phase-ring" that is fixed within the back focal plane of the objective is needed. The "phase-ring" has two important tasks:

1. It has to achieve an alignment of the brightness of refracted and unrefracted light as the rays that pass through the preparation are weakened in their intensity. In contrast to a conventional image gained with the help of a light microscope, the background of a phase-contrast image is thus dark.
2. The phase difference of most biological preparations is one-quarter of a wavelength or less (*lambda*/4). The phase-ring is constructed so as to achieve an additional difference of *lambda*/4. The complete phase difference is thus *lambda*/2 and crest and trough of refracted and unrefracted rays extinguish each other. A disadvantage of this method is the appearance of light halos around some objects ("halo-effect").

The main advantage is live cell can be seen without staining

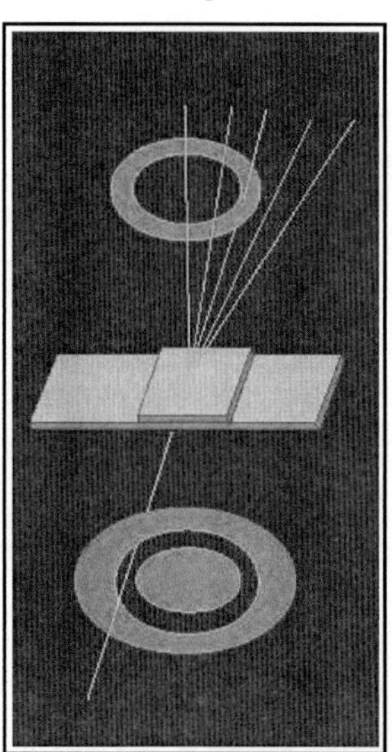

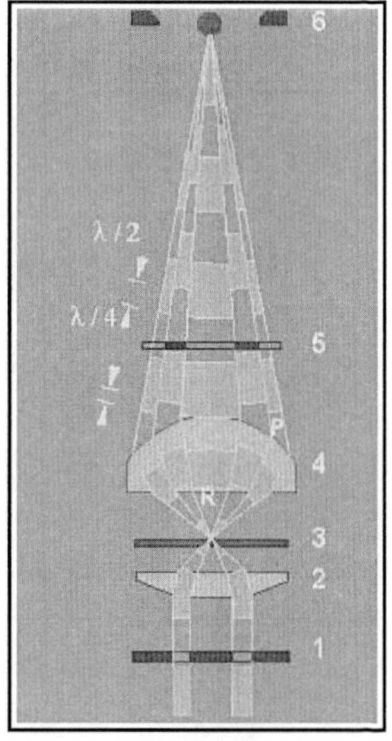


*To the left:* Arrangement of the ring-shaped mask below the objective and of the phase-ring within the objective. Only the direct light beams are influenced by the phase-ring. *To the right:* The path of light rays within a phase-contrast microscope. 1. ring-shaped mask, 2. condenser, 3. specimen, 4. objective, 5. phase plate, 6. focal plane of the objective. The wave character of the light is indicated by the change of light and dark areas.

## Dark Field Microscopy

This method uses a special condenser with a big aperture through which light beam passes through and also the objective. Only if the object is brought into the center of the light, the light is diffracted, collected by the objective, and used for image formation. Shiny structures are seen in front of a dark background. This method has no very important role in biology, but is impressive with crystals.

## Fluorescence Microscopy

Fluorescence microscopy is based on the fact that some molecules emit part of the light absorbed by them as longer waves. A well-known example is the red fluorescence of chlorophyll. Since the Austrian teacher M. Haitinger made his examinations in the thirties of this 20th century it is known that a number of so-called fluorochromes exist, with which microscopic preparations can be stained so that they emit fluorescence indirectly. In the following decades, a whole range of so-called vital dyes were detected or developed, that -used in low concentrations- could mark specific parts of living cells or of tissues. They enabled researchers to follow the solute transport within cells and tissues or to ascertain the pH of special compartments, like for example the vacuole.

Since roughly twenty years fluorescence microscopy is flowering again, on one hand, because the range of fluorochromes became much broader and on the other hand, because completely new approaches like indirect fluorescence could be used successfully. Additionally the construction of microscopes was improved and highly efficient masks were developed.

There are two ways in which fluorescence microscopes can be constructed: as *epifluorescence*- and as *transmission microscopes*. The second is the older way of construction. Three components are needed:

1. A strong source of light, that emits mainly short lightwaves. Mercury high pressure lights have proven useful.
2. *The first barrier filter:* This filter helps to shut off all radiation other than the one that activates the specific dye. It is placed behind the light source within the light cone. Besides it is advantageous to work with a dark field condenser.
3. *Second barrier filter:* This filter is brought into the light cone between objective and eyepiece. It lets through only long wavelengths, that are caused by emission of the preparation (so-called "secondary radiation").

In the last years the *epifluorescence microscope* has replaced the transmission fluorescence microscope more and more. But the transmission fluorescence microscope is still better suited to only weakly magnifying objectives (2.5 x, 6.3 x). With the epifluorescence the objective is also a condenser and the stronger it is the more intensive radiation can be used. The heart of epifluorescence is a construction within the light cone between objective and eyepiece that serves to feed activating radiation into the system and is constructed of first barrier filter, beam-splitting mirror and second barrier filter.

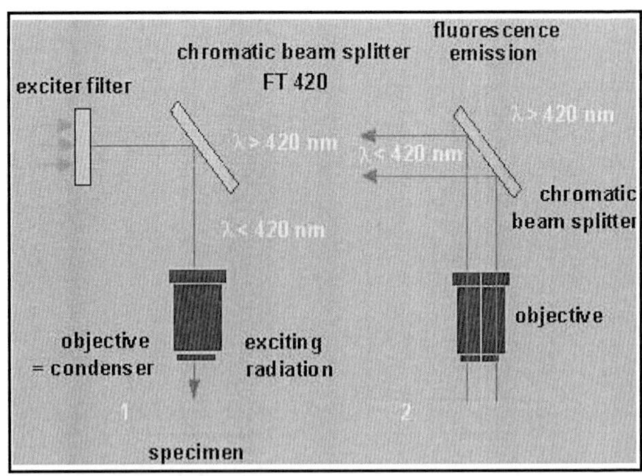

Excitation and fluorescence with chromatic beam splitters. Similar to the interference filters these are specially coated mirrors used under 45° to the illuminating beam. They reflect certain spectral ranges, while others are completely transmitted. The separating line between reflection and transmission may be set at any point of the spectrum. 1. Exciting radiation 2. Fluorescence emission.

## Electron Microscopy

In 1924 L. de Broglie discovered the wave-character of electron rays thus giving the prerequisite for the construction of the electron microscope. The prototype was built by M. Knoll and E. Ruska (Technische Universität Berlin, 1932). One of the first biological objects depicted was the tobacco mosaic virus (TMV). The first picture of a cell was published in 1945 by K.R. Porter, A. Claude and E.F. Fullam (Rockefeller Institute, New York).

## The Transmission Electron Microscope (TEM)

The conventional electron microscopy is nowadays called TEM (transmission electron microscopy). We will therefore start with its construction. The ray of electrons is produced by a pin-shaped cathode heated up by current. The electrons are vacuumed up by a high voltage at the anode. The acceleration voltage is between 50 and 150 kV. The higher it is, the shorter are the electron waves and the higher is the power of resolution. But this factor is hardly ever limiting. The power of resolution of electron microscopy is usually restrained by the quality of the lens-systems and especially by the technique with which the preparation has been achieved. Modern gadgets have powers of resolution that range from 0.2-0.3 nm. The useful resolution is therefore around 300,000 x.

The accelerated ray of electrons passes a drill-hole at the bottom of the anode. Its following way is analogous to that of a ray of light in a light microscope. The lens-systems

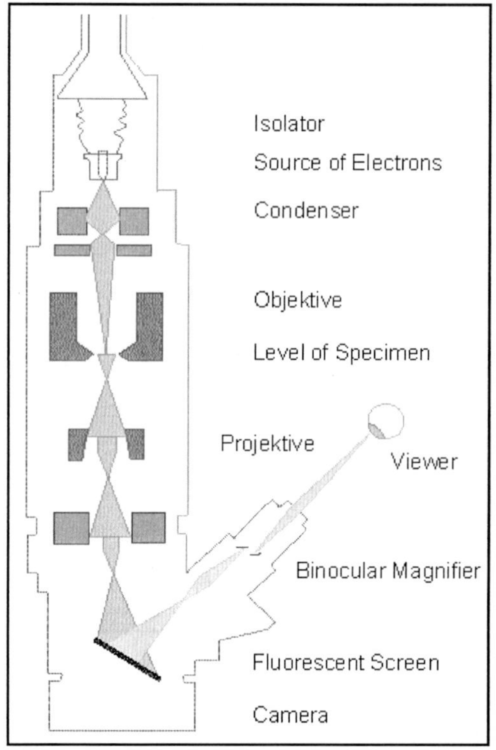


Transmission Electron Microscope (TEM)

consist of electronic coils generating an electromagnetic field. The ray is first focused by a condenser. It then passes through the object, where it is partially deflected. The degree of deflection depends on the electron density of the object. The greater the mass of the atoms, the greater is the degree of deflection. Biological objects have only weak contrasts since they consist mainly of atoms with low atomic numbers (C, H, N, O). Consequently it is necessary to treat the preparations with special contrast enhancing chemicals (heavy metals) to get at least some contrast. Additionally they are not to be thicker than 100 nm, because the temperature is raising due to electron absorption. This again can lead to destruction of the preparation. It is generally impossible to examine living objects.

After passing the object the scattered electrons are collected by an objective. Thereby an image is formed, that is subsequently enlarged by an additional lens-system (called projective with electron microscopes). Thus formed image is made visible on a fluorescent screen or it is documented on photographic material. Photos taken with electron microscopes are always black and white. The degree of darkness corresponds to the electron density (= differences in atom masses) of the candled preparation.

## The Scanning Electron Microscope (SEM)

The path of the electron beam within the scanning electron microscope differs from that of the TEM. The technology used is based on television techniques. The method is suitable for the depiction of preparations with conductive surfaces. Biological objects have thus to be made conductive by coating with a thin layer of heavy metal (usually gold is taken). The power of resolution is normally smaller than in transmission electron microscopes, but the depth of focus is several orders of magnitude greater. Scanning electron microscopy is therefore also well-suited for very low magnifications.

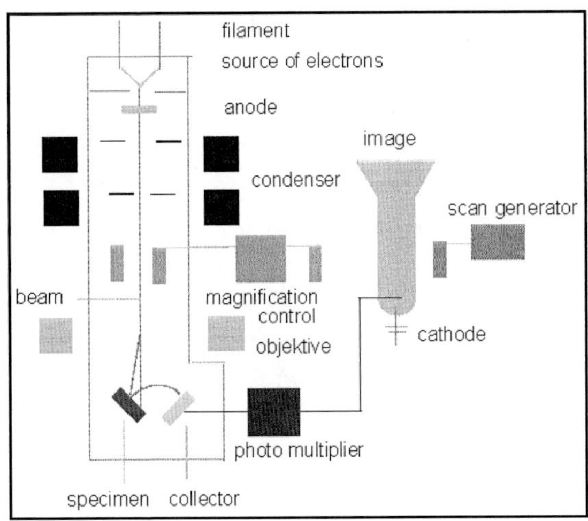


Scanning Electron Microscope (SEM)

The surface of the object is scanned with the electron beam point by point whereby secondary electrons are set free. The intensity of this secondary radiation is dependent on the angle of inclination of the object's surface. The secondary electrons are collected by a detector that sits at an angle at the side above the object. The signal is then enhanced electronically. The magnification can be chosen smoothly (depending on the model) and the image appears a little later on a viewing screen.Finally some outlines of new and further developments are given.

1. *The high voltage electron microscope:* It operates with an accelerating voltage of 700 - 3000 kV. Its power of resolution is greater, the preparation can be thicker, the strain on the preparation is smaller. But the enormous technical expenditure is disadvantageous. Only few gadgets exist. New results concerning botany have not been gained.

2. *The scanning transmission electron microscope (STEM):* In this development of the SEM do the electrons pass through the preparation and the secondary radiation thus generated is used for image formation. Here, too, the expenditure is large, but it is still worthwhile, since large molecules like nucleic acids or proteins or molecular complexes like viruses can be depicted much better and gentler than with the TEM. No news for botany, though.

The interpretation of images gained with electron microscopy is increasingly done with computerized interpretation programs. But they are usually only suitable for the reconstruction of regularly recurring patterns and these, again, are found more often on a molecular level than on a cellular one.

CHAPTER 4

# Groups of Microorganisms

## Place of Microorganisms in Living World

*Haeckel's kingdom Protista*

E.H. Haeckel, 1866 proposed this concept. Apart from plant and animal kingdom, the third kingdom "*Protista*" was developed by him. The protista includes unicellular organisms that are typically neither plant nor animal. These organisms, the protists, include bacteria, algae, fungi and protozoa. Bacteria are referred as lower protists and others are called as higher protists.

*Whittaker's five kingdom concept*

R.H.Whittaker (1969) proposed the concept of five kingdom. He divided the living things in to 5 kingdoms.

1. Planta (photosynthetic)
2. Fungi (Nutrient absorption)
3. Animalia (nutrient ingestion)
4. Protista (Eukaryotic, unicellular) includes microalgae and protozoa
5. Monera (prokaryotic) includes bacteria.

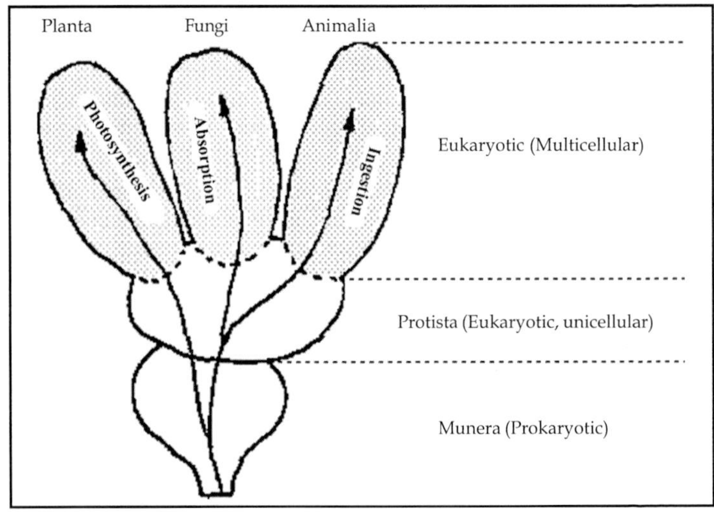


Wittaker's five kingdom model

*Three domains of life (Present concept)*

Based on the molecular taxonomy, the living organisms were divided into 3 domains. (*Domain* refers the highest level of classification like kingdom)

Recent work, however, has shown that what were once called "prokaryotes" are far more diverse than anyone had suspected. The Prokaryotae are now divided into two domains, the Bacteria and the *Archaea*, as different from each other as either is from the *Eukaryota*, or eukaryotes. No one of these groups is ancestral to the others, and each shares certain features with the others as well as having unique characteristics of its own.

Domain 1. *Bacteria* - includes prokaryotic organisms

Domain 2. *Archaea* - includes certain prokaryotes found in enormous numbers, often from unusual habitats such as very hot, very salty, or very anaerobic (no oxygen)

Domain 3. *Eukaryota* – including eukaryotic organisms

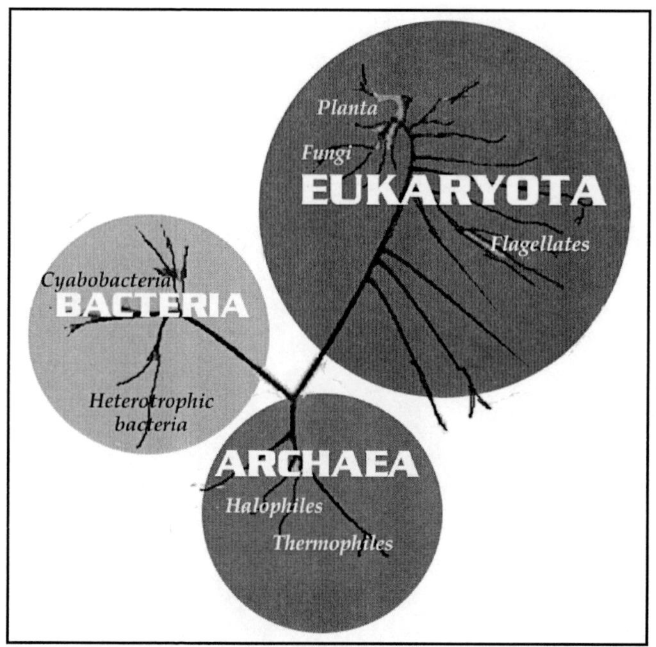


Three domain of life

*Bergey's manual of systematic bacteriology*

Bergey's manual is the international standard for bacterial taxonomy. The entire world of bacteria and archaea were divided into 2 kingdoms. Kingom I is *Archaeota* and the Kingdom II is *Bacteria*. The manual divide the bacteria into 5 volumes. Volume I include archaea, cyanobacteria and phototrophs Volume II include the proteobacteria Volume III-Bacteria with low G+C Gram positives and Volume IV-high G+C Gram positives and Volume V-Planctomycetes, spirochaetes, Fibrobacteres, Bacteroides and Fusobacteria.

*Prokaryotes* : The organism *lacking* true nucleus (membrane enclosed chromosome and nucleolus) and other organelles like mitochondria, golgi body, endoplasmic reticulum etc. are referred as Prokaryotes. (Ex : Bacteria, Archaea).

*Eukaryotes* : The organism possessing membrane enclosed nucleus and other cell organelles are referred as Eukaryotes (*Ex* : algae, fungi, protozoa).

| Characteristic | Prokaryotes | Eukaryotes |
|---|---|---|
| Diagrammatic representation | 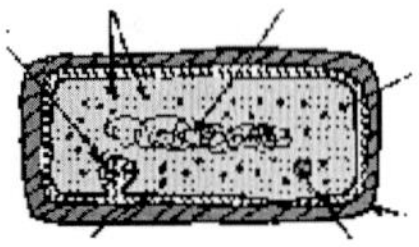 | 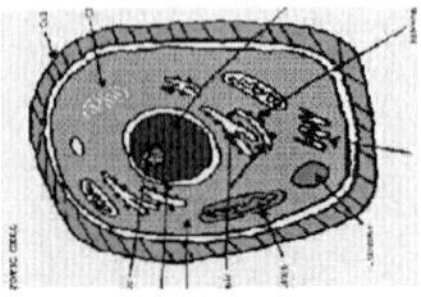 |
| Electron micrograph | 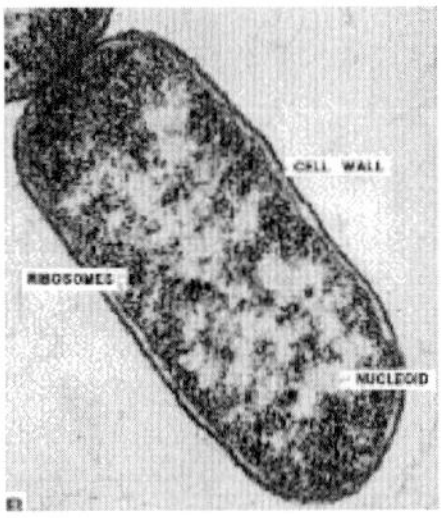 | 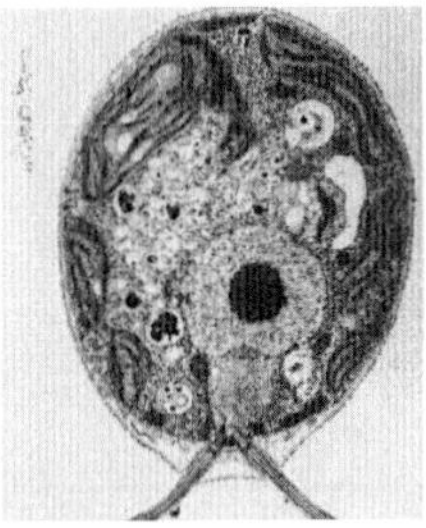 |
| Size of cell | Typically 0.2-2.0 mm in diameter | Typically 10-100 mm in diameter |
| Nucleus | No nuclear membrane or nucleoli (nucleoid) | True nucleus, consisting of nuclear membrane & nucleoli |
| Membrane-enclosed organelles | Absent | Present; examples include lysosomes, Golgi complex, endoplasmic reticulum, mitochondria & chloroplasts |
| Flagella | Consist of two protein building blocks | Complex; consist of multiple microtubules |

*Contd...*

| | | |
|---|---|---|
| Glycocalyx | Present as a capsule or slime layer | Present in some cells that lack a cell wall |
| Cell wall | Usually present; chemically complex (typical bacterial cell wall includes peptidoglycan) | When present, chemically simple |
| Plasma membrane | No carbohydrates and generally lacks sterols | Sterols and carbohydrates that serve as receptors present |
| Cytoplasm | No cytoskeleton or cytoplasmic streaming | Cytoskeleton; cytoplasmic streaming |
| Ribosomes | Smaller size (70S) | Larger size (80S); smaller size (70S) in organelles |
| Mesosomes | Present | Absent |
| Chromosome arrangement | Single circular chromosome; lacks histones | Multiple linear chromosomes with histones |
| Extrachromosomal DNA | Plasmid | Mitochondria and chloroplast |
| Cell division | Binary fission | Mitosis |
| Sexual reproduction | No meiosis; transfer of DNA fragments only (conjugation) | Involves meiosis |
| Site for cellular respiration | Cell membrane | Mitochondria |
| Locomotion | Rotating flagella and gliding | Undulating flagella and cilia, and also amoeboid movement |
| Pili | Sex or attachment pili present | absent |

## Groups of Microorganisms

The microorganisms are divided into six distinct groups. There descriptions are as follow :

a. *Bacteria :* are group of phylogenetically related prokaryotic organisms distinct from archaea.

- size ranges from 0.5 - 3.0 mm.
- They are unicellular.
- presence of cell wall (composed of lipopolysaccharide and peptidoglycon).
- simple internal structures.
- able to grow in artificial media.
- reproduce asexual simple cell division (binary fission).

*Importance*

- some cause disease.
- important role in nutrient cycling in soil.
- useful in industrial production of antibiotics, vitamins, dairy products, value added products.

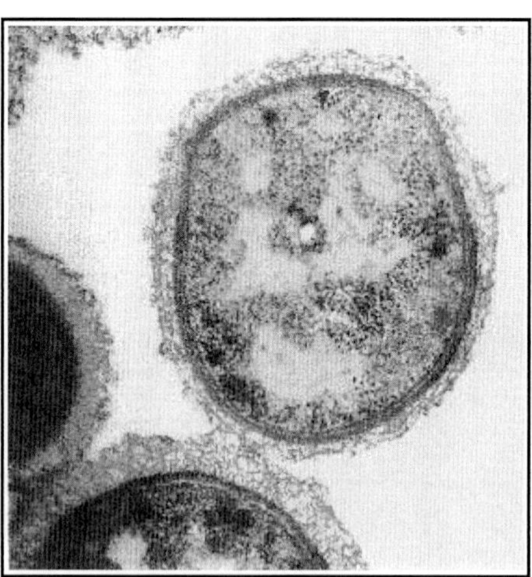

Pseudomonas aeruginosa

b. *Archaea* : are group of phylogenetically related group of prokaryotic microorganisms which are primitive and differ with bacteria

- size ranges from 0.5 - 2.5 mm.
- They are unicellular.
- simple internal structures.
- differ with bacteria by absence of peptidoglycon in cell wall.
- Most of them live in extreme environment (high salt, methane producing organisms, thermophiles).

*Importance*

- Methane (biogas) production;
- Waste water treatment and energy production;
- Temperature resistant enzymes for molecular biology and biochemistry (ex. DNA polymerase from *Thermococcus*).

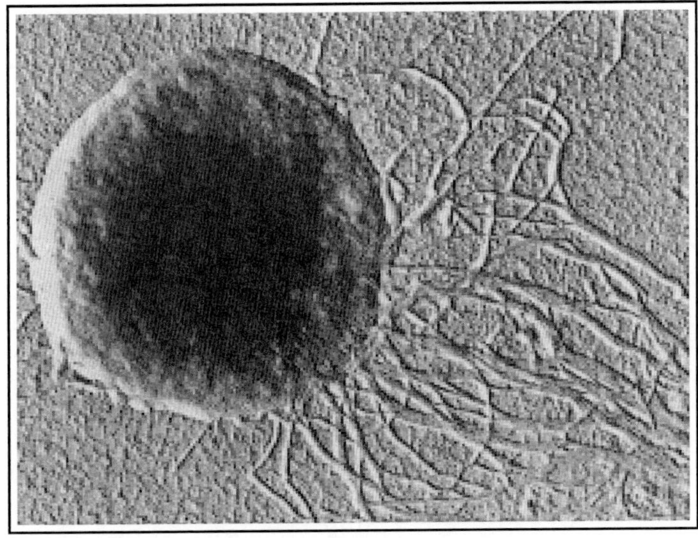

*Methanococcus* sp

C. *Fungi* : are the group of eukaryotic microorganisms lack of chlorophyll.

- size ranges from 5 - 10 mm
- consist of both unicellular (yeast) and multicellular (mushroom) organisms
- saprophytic nature (derive food through dead organic matter) (renamed as *organotrophs*)
- Presence of cell wall (made of *Chitin* - polymer of N-acetyl glucosamine)
- Presence of asexual and sexual reproduction.

*Importance*

- Cause plant diseases; few animal and human diseases too.
- Decomposition of organic matter in the soil and nutrient cycle.
- Food spoilage.
- Yeast - industrial importance.
- Antibiotic industry.
- Used as food (mushroom).

Mushroom Fungus

*D. Protozoa (protos – first; zoon – animal)* : are group of eukaryotic microorganisms lack of cell wall.

- size ranges from 2.0 to 200 mm.
- Large numbers are survive in soil, fresh water and marine water.
- Most are motile.
- Reproduction by asexual and sexual processes.

*Importance*

- They play important role in food chain of aquatic environment.
- Some cause diseases.
- Importance in the ecological balance of many microbial communities in soil and aquatic environment.
- Industrial waste treatment.
- Sewage water treatment.

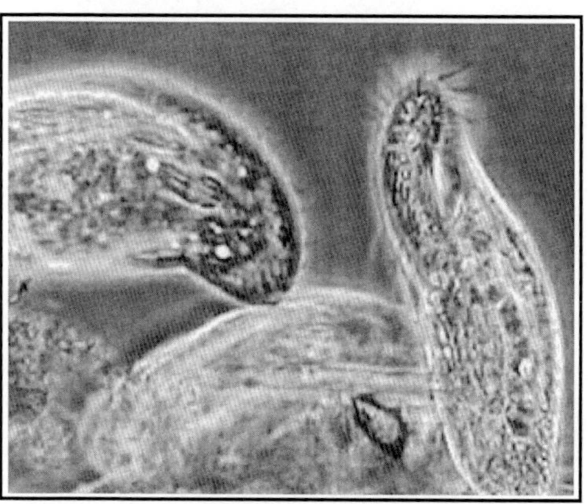

*Euglena*

*E. Algae* : are the group of eukaryotic microorganisms that contain chlorophyll and carry out the oxygenic photosynthesis.

- contain different type of pigments.
- size ranges from 1 mm to many meters.
- Unicellular and multicellular.
- Most occur in aquatic environment.
- reproduction by sexual and asexual process.

*Importance*

- Food producer of aquatic environment (primary producer)
- Used as food supplements
- Used in pharmaceutical industry
- Economically important products - agar, alginic acid, carrageenan
- Some cause disease

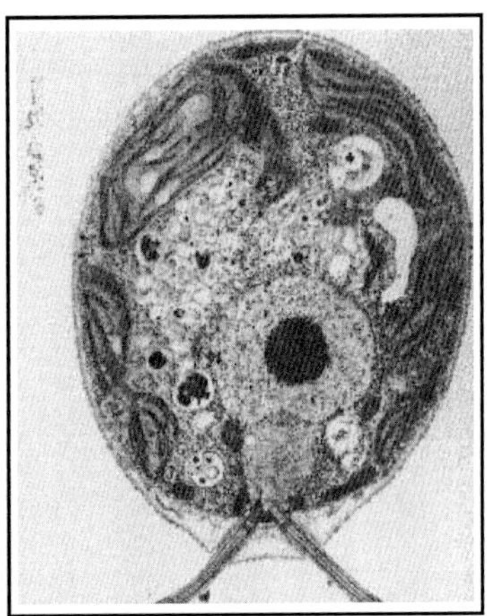

*Chlamydomonas*

F. *Viruses* : are non cellular microorganisms.

- Size ranges from 0.015 - 0.2 mm.
- Cannot grow in laboratory condition (needs living host).

- Obligate parasites.
- Lack metabolic machinery for energy generation and reproductive system for replication.
- Depends higher organisms for energy and reproduction.

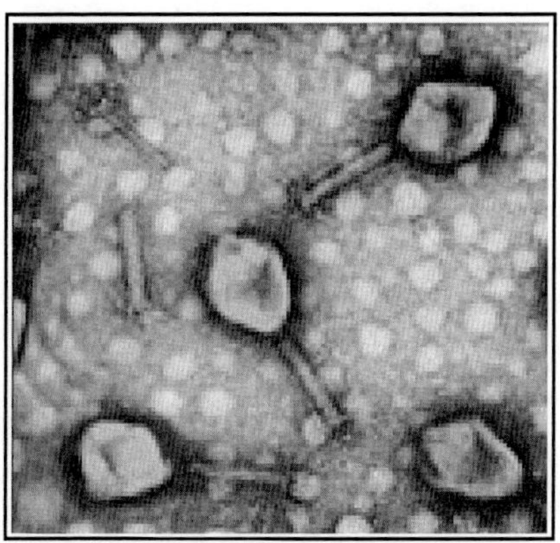

Bacteriophage

*Importance*

- Cause diseases to human, animal, microorganism.
- Few used in molecular biology as vector (to transfer the genetic elements to an organism).

CHAPTER

5

# Bacteria

## Size - Shapes - Arrangements - Morphology

Bacteria are very small, most being approximately 0.5 to 1.0 mm diameter. Smallest bacteria are *Mycoplasmas*, as small as 0.2 micrometers (almost as small as largest poxviruses).

Accepted wisdom is that bacteria are smaller than eukaryotes. But certain cyanobacteria are quite large; *Oscillatoria* cells are 7 micrometers diameter, size of red blood cells. And certain eukaryotes (e.g. *Nanochlorum eukaryotum*) are very small, only 1 to 2 micrometers, but true eukaryotes (nucleus, chloroplast, mitochondrion are present). So size difference is, like many generalizations, only a useful yardstick, not an absolute truth.

*Epulopiscium fishelsoni*, discovered in 1985 in intestinal tract of sturgeonfish, is an enormous, cigar-shaped cell, as large as 80 x 600 micrometers (that's 0.6 mm, large enough to be seen by the naked eye). Amazingly, this cell is prokaryotic! Initial evidence by EM was hard to believe, but confirmed rRNA comparisons with other organisms, a cousin of Gram-positive *Clostridium* genus.

The relative size of the bacteria are as follows :

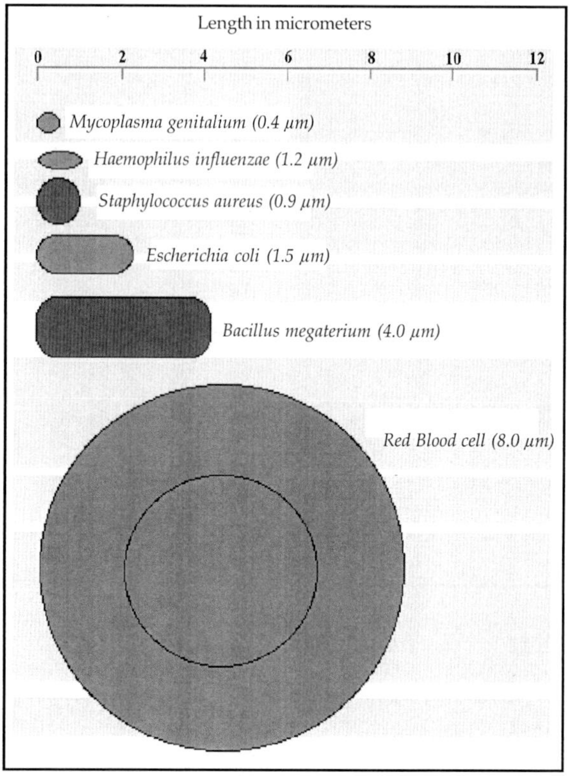

### *Shape*

The shape of the bacterium is governed by its rigid cell wall. Typical bacterial cells are spherical (Cocci and Coccus) or straight rod shaped (Bacilli and Bacillus) or helically curved rods (Spirilli and Spirillum). Few bacteria can exhibit variety of shapes that are referred as Pleomorphic. (Ex. *Arthrobacter*).

Example for cocci – *Streptococcus*

Example for rods – *Bacillus*

Example for curved rods - *Azospirillum*

Cocci appear in several characteristic arrangements, depending on the plane of cellular division and whether the daughter cells stay together or not.

Spherical is called *coccus*.

Division along the same plane forms chains; 2 cocci together - *Diplococcus*.

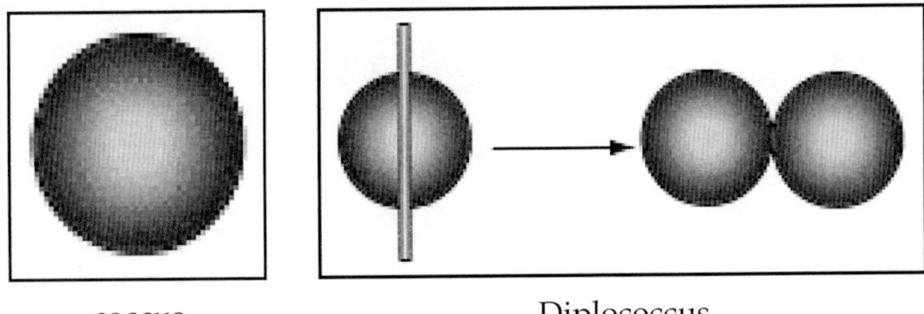

coccus Diplococcus

4 - 20 in chains - *Streptococcus*.

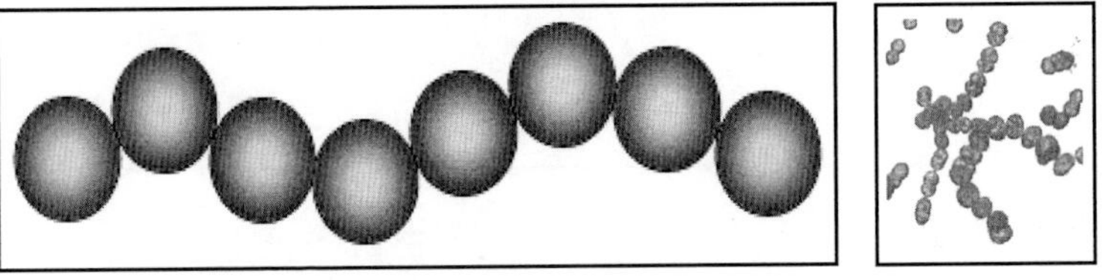

Streptococcus

Division along 2 different planes - *Tetrads or Tetracoccus*.

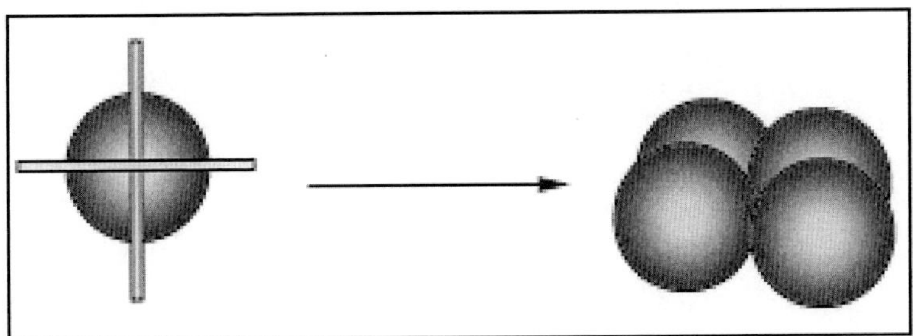

Tetrads or Tetracoccus

Division along 3 planes regularly - *Sarcina*

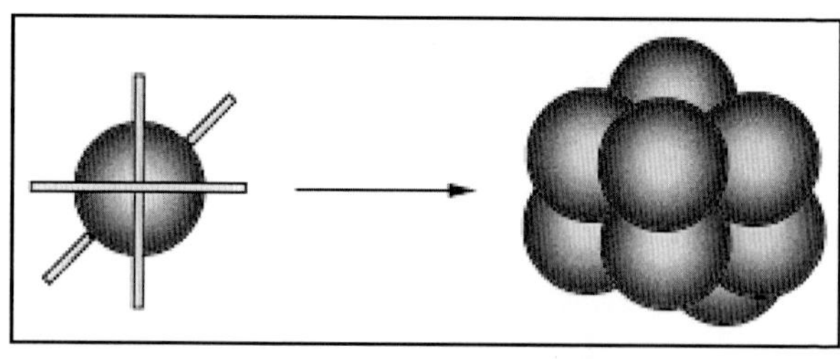

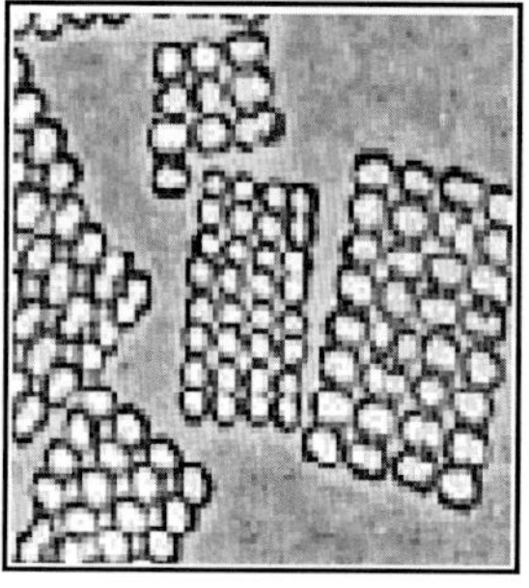

Sarcina

Division along 3 planes irregularly – *Staphylococcus*

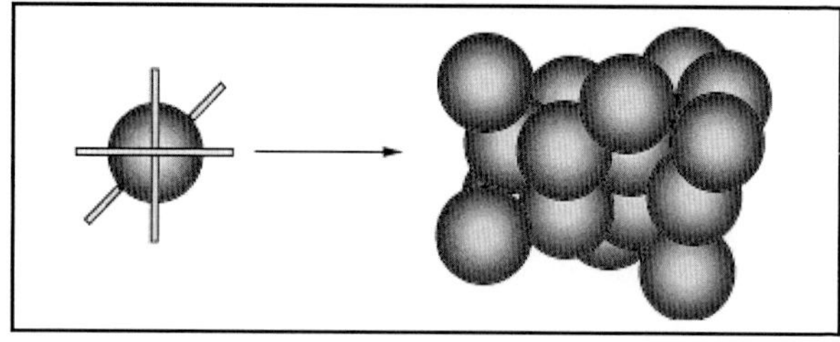

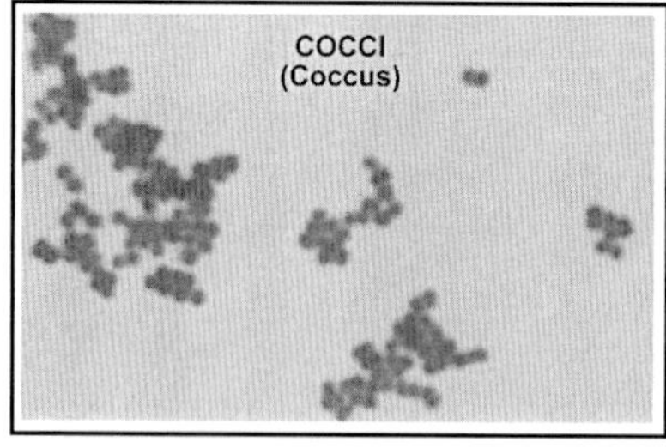


Staphylococcus

Bacilli are not arranged in patterns as complex as those of cocci and mostly occur in single or in pairs. The following are some arrangements of rod shaped bacteria.

Rod shape is called *Bacillus*.

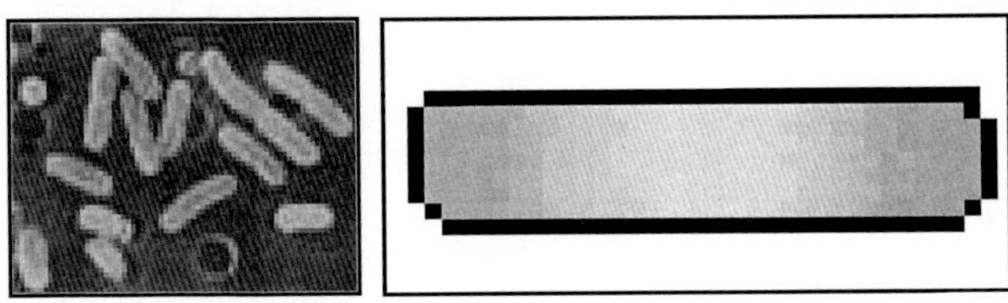

Bacillus

Two bacilli together - *Diplobacilli*

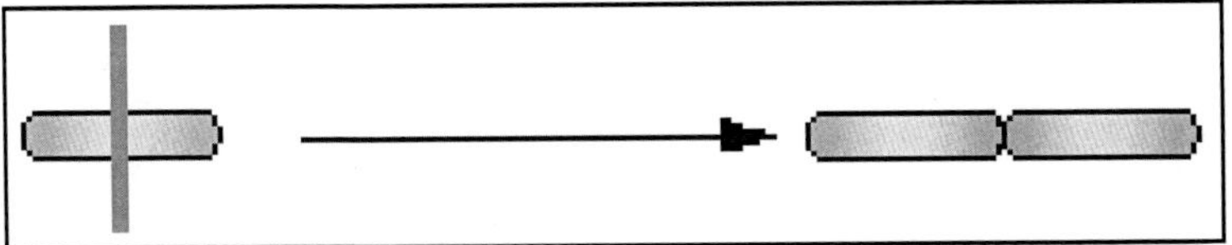

Diplobacilli

Chains of bacilli are called *Streptobacilli*

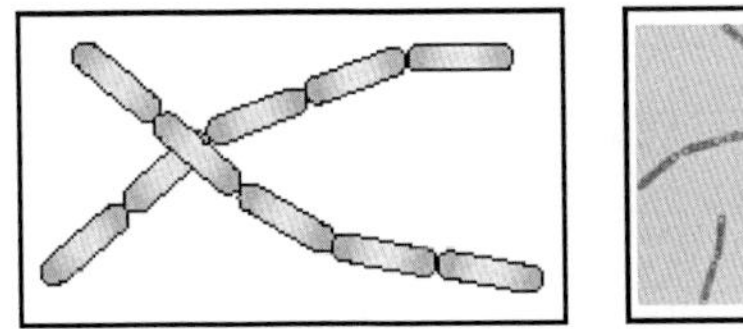

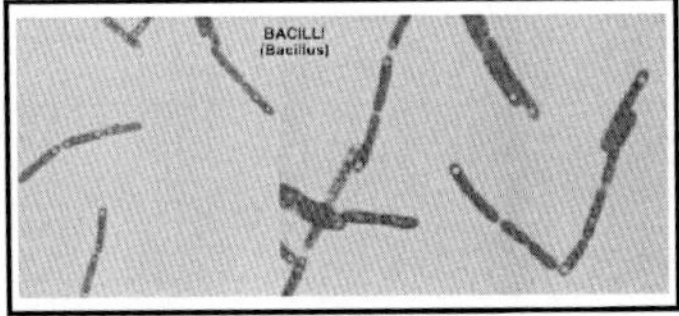


Streptobacilli

*Palisades :* Rods side by side arrangement like match sticks or in X, V or Y figures. Ex. *Corynebacterium diphtheria*

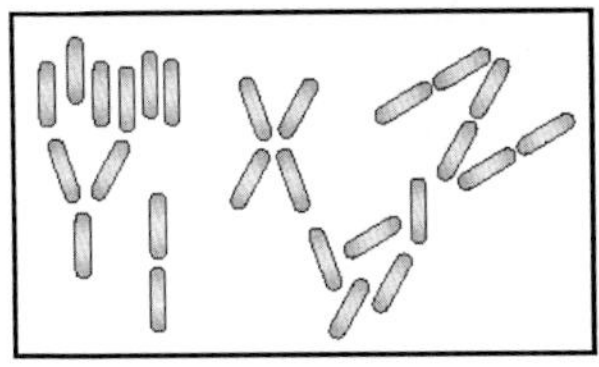

Palisades

Curved bacteria (spirilla) vary with the number of twists they had.

Spiral shape that is rigid is called *Spirillium.*

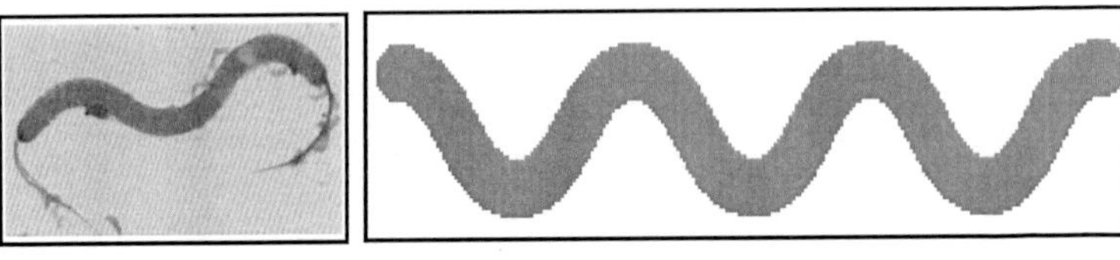

Spirillium

Spirals with less than one complete twist are called as *Vibrioid* shapes (Ex. *Vibrio*).

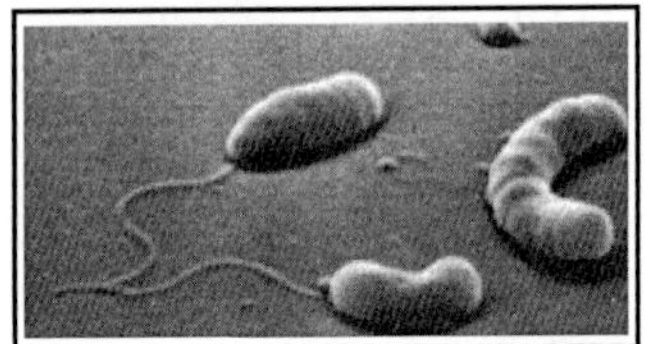

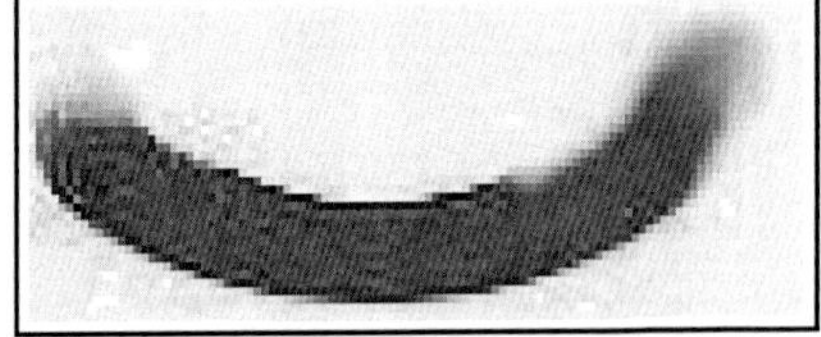

Vibrioid

If the organism is flexible and undulating, it is called *Spirochete.*

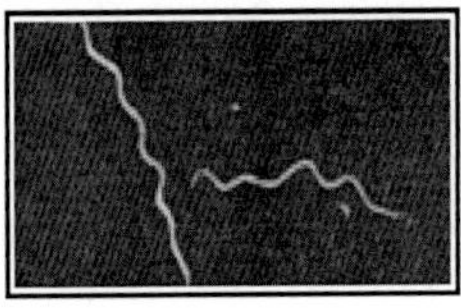

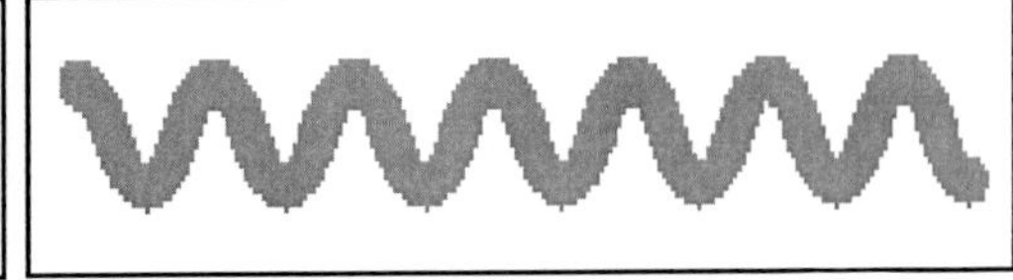

Spirochete

*Other shapes*

Filamentous shaped – *Streptomyces* sp. form long, multinucleated, branched filaments called as hyphae

Pear shaped – Ex. *Pasteuria*

Lobed spheres - Ex. *Sulfolobus*

Rod with square end – Ex. *Bacillus anthracis*

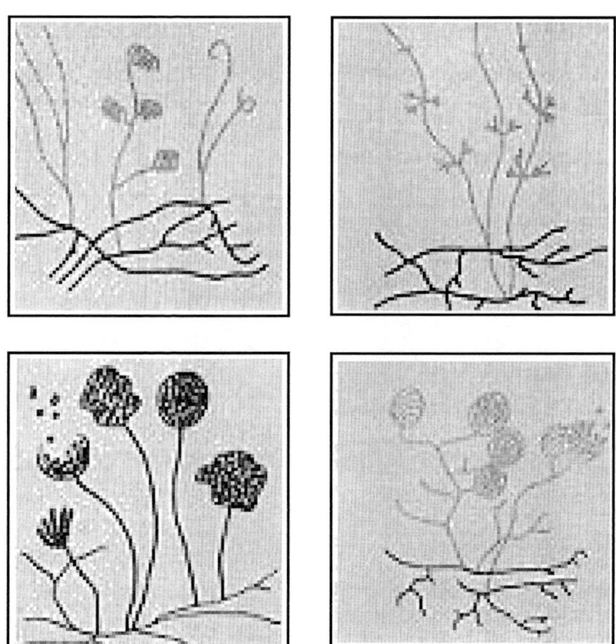

Actinomycetes - hyphae

## Bacteria – Morphology

Bacteria are unicellular organisms of relatively simple construction, especially if compared to eukaryotes. Whereas eukaryotic cells have a preponderance of organelles with

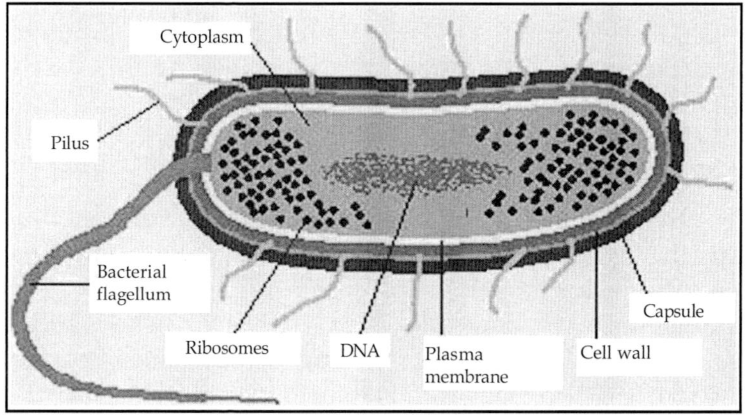

separate cellular functions. Prokaryotes carry out all cellular functions as individual units.

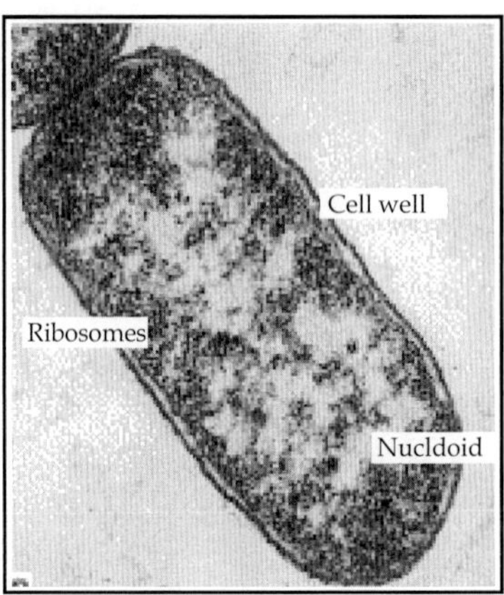


Bacteria – Morphology

A prokaryotic cell has five essential structural components: a genome (DNA), ribosomes, cell membrane, cell wall and a surface layer.

Other than enzymatic reactions, all the cellular reactions incidental to life can be traced back to the activities of these macromolecular structural components. Thus, functional aspects of prokaryotic cells are related directly to the structure and organization of the macromolecules in their cell make-up, i.e., DNA, RNA, phospholipids, proteins and polysaccharides. Diversity within the primary structure of these molecules accounts for the diversity that exists among bacteria.

### *1. Flagella (Sin. flagellum)*

*Flagella (Sin. flagellum) :* are hair like, filamentous helical appendages that protrude through cell wall which is responsible for movement of the bacterium. They are very thinner than the eukaryotic flagellum. The diameter is about 20 nm, well below the resolving power of compound microscope. The flagellum is rotated by a motor apparatus in the plasma membrane allowing the cell to swim in the fluid environment.

Flagellum is composed of three parts namely, *basal body* (associated with cytoplasmic membrane and cell wall), *hook* (next to basal body) and a *filament* (a hair like, longer than the cell). The flagellum is made up of *flagellin* protein.

The flagellum is powered by the proton motive force of the cell membrane. About half of the rod shaped, all the spiral and curved and very few spherical bacteria are motile by flagella.

The distribution/ arrangement of flagellum in the bacteria vary with different types.

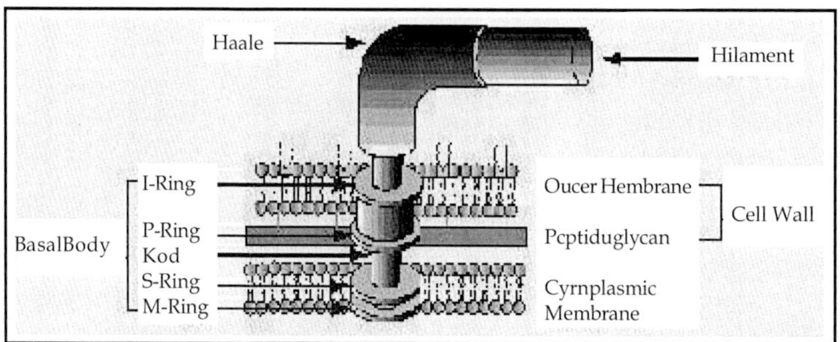


Flagella Structure in Gram Negative Organism

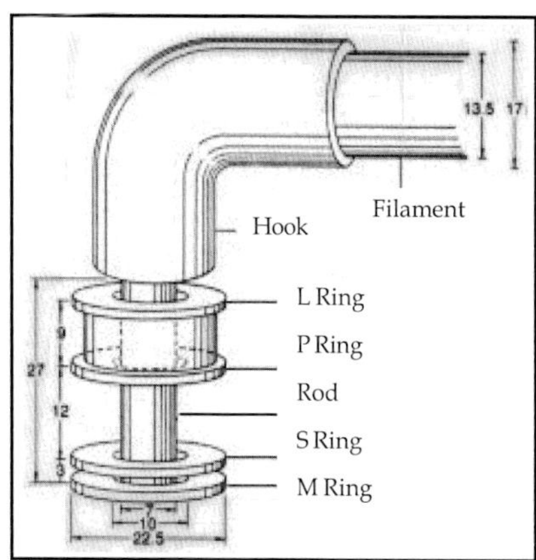


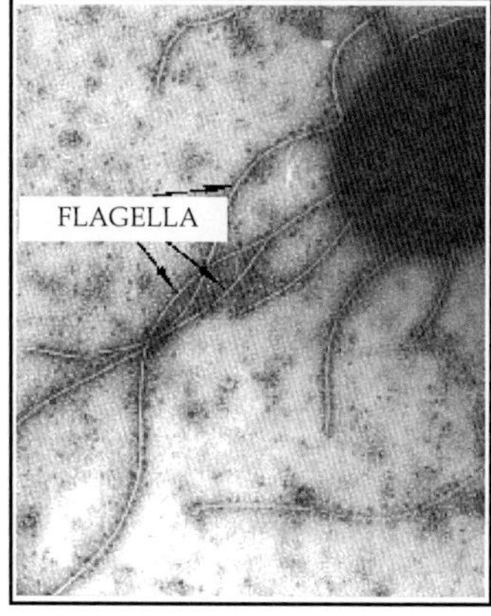

*Diagramatic Representation of Flagella*

a. *Monotrichous flagellum :* single flagellum present in the pole of the bacterium (Ex. *Pseudomonas aeruginosa*).

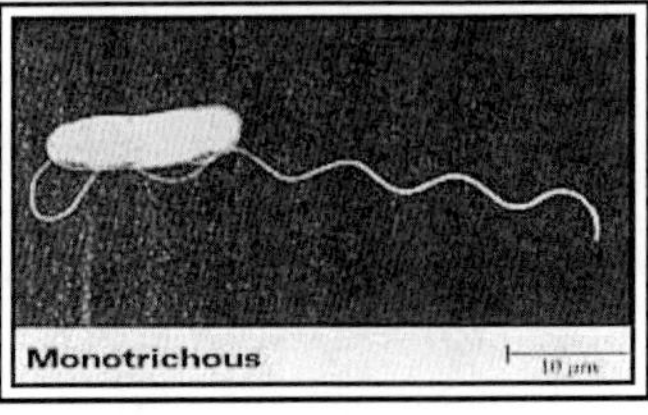


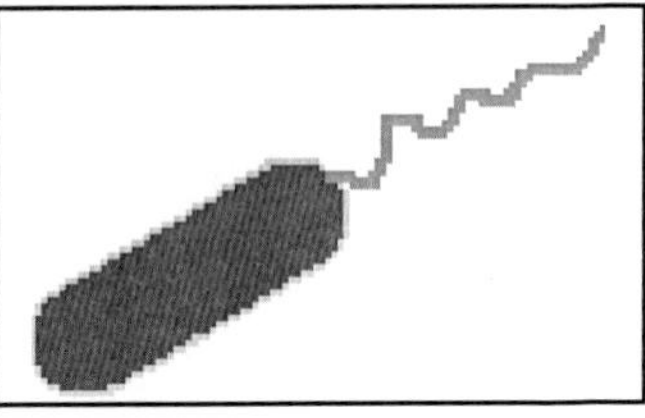

b. *Lophotrichous flagella :* cluster of flagella present in the pole of the bacterium (*Ex. Pseudomonas fluorescens*).

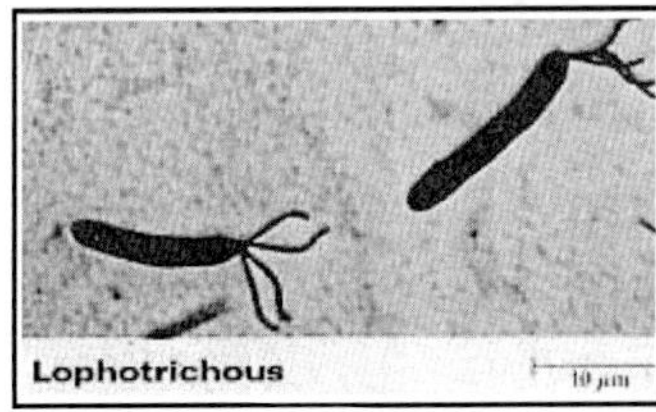


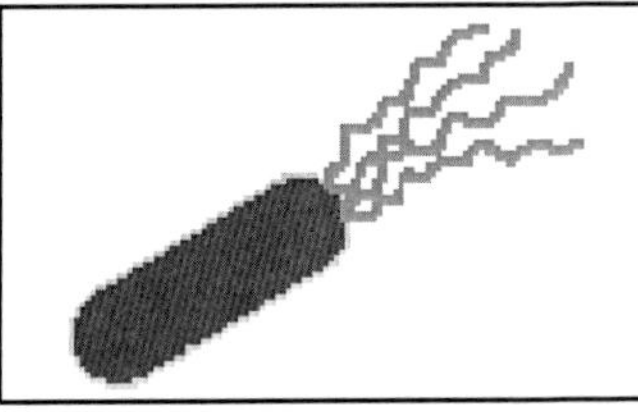

c. *Amphitrichous flagella :* single or cluster of flagella present in the both pole of bacterium (Ex . *Aquaspirillum serpens*).

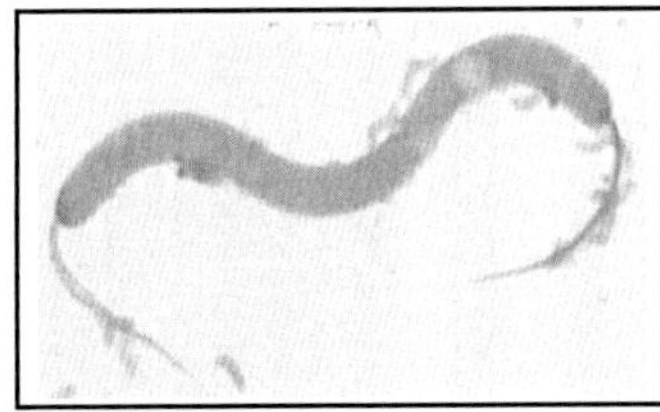

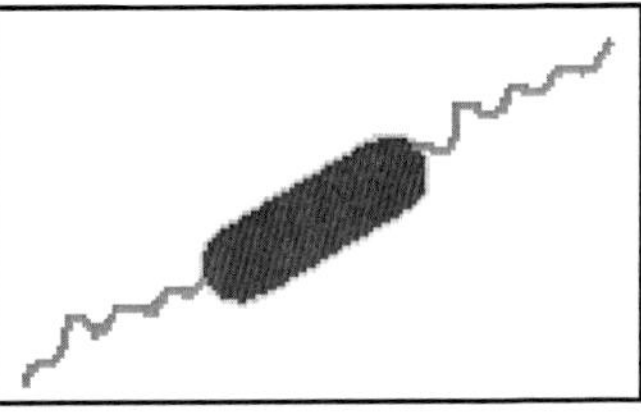

d. *Peritrichous flagella :* flagella surrounded through out the body of the bacterium (Ex. *Salmonella typhi*).

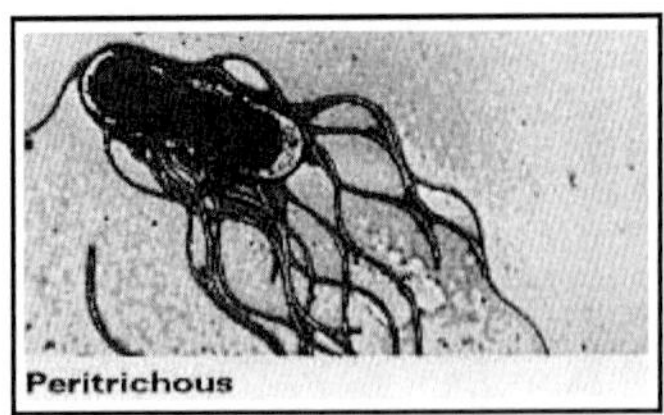


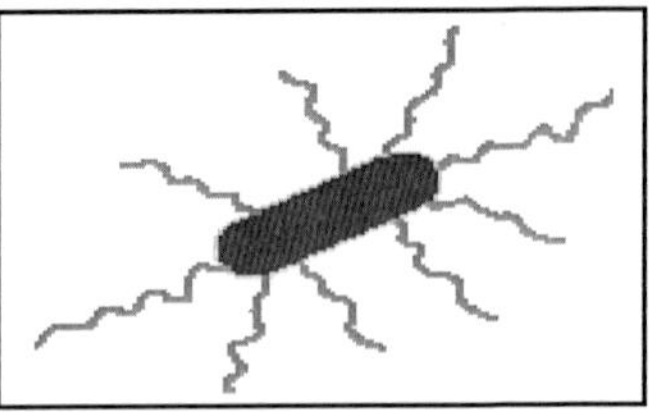

*Endoflagella*

The flagella or filaments, present between the outer membrane and inner membrane of the cell wall of the bacterium are called as *endoflagella* or *periplasmic flagella* or *axial filaments*. They are attached at one end of the cell. Ex. Spirochetes- group of bacteria; Bacterium : *Spirochete, Leptospira.*

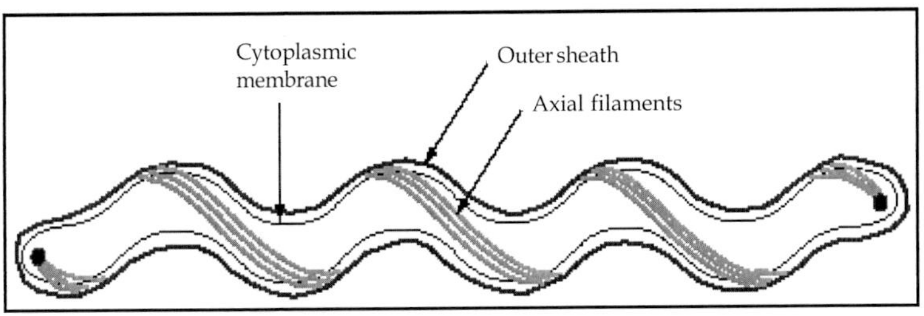


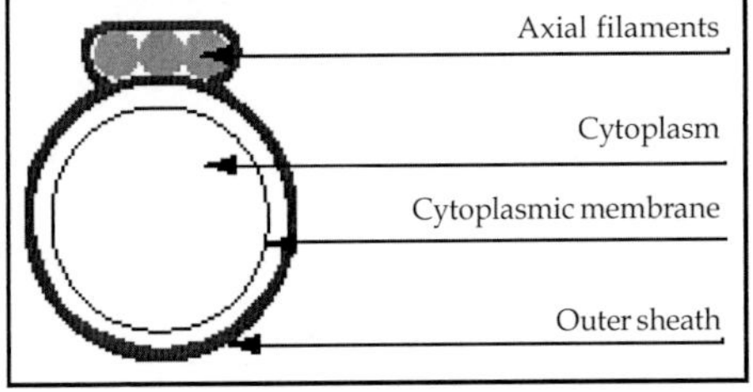


The flagella are responsible for motility of the bacterium. It rotates like a propeller. Rings in the basal body rotate relative to each other causing the flagella to turn. The energy to drive the basal body is obtained from the proton motive force. The average speed of bacterial movement is 50 mm/sec which equivalent to its 10 body length.

*2. Pili (Sin: pilus)*

Pili are hollow, non-helical, filamentous appendages thinner, shorter and more numerous than flagella. They also present in the non-motile bacteria too. Like flagella, they composed of protein, called pilin. They also called as *Fimbriae*. They are very common in gram –ve bacteria, but occur in some archaea and gram +ve bacteria too.

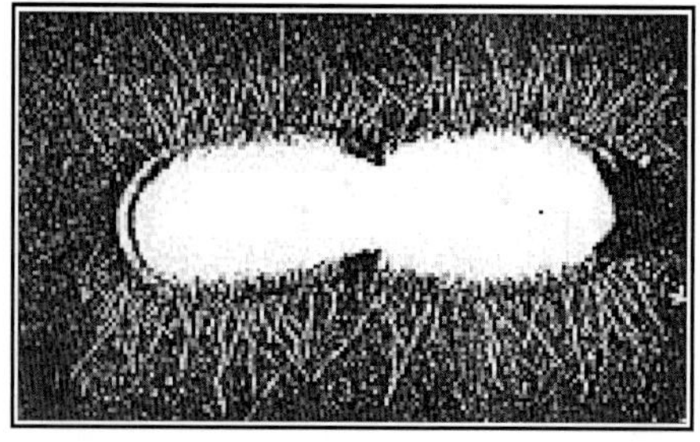

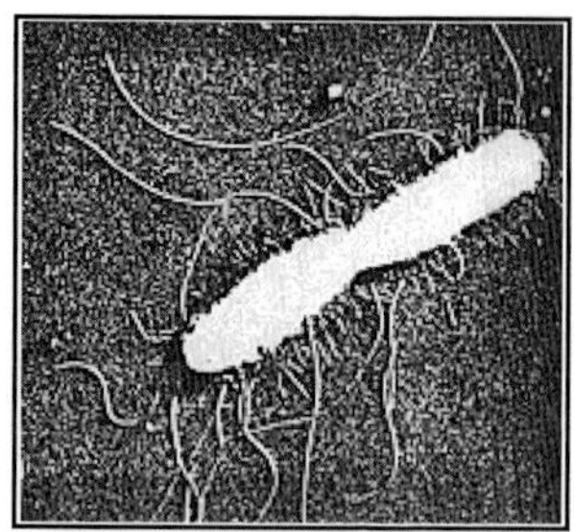

Pili are mainly involved in the *adherence or attachment* of bacteria to surfaces, substrates and other cells in nature.

*Some pili (F or sex pili)* are involved in the transfer of DNA from one cell to other, called *conjugation*. In *E. coli*, both sex pili and common pili are present).

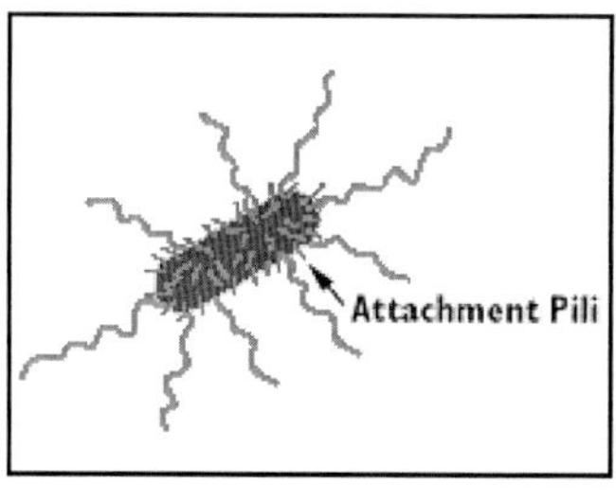


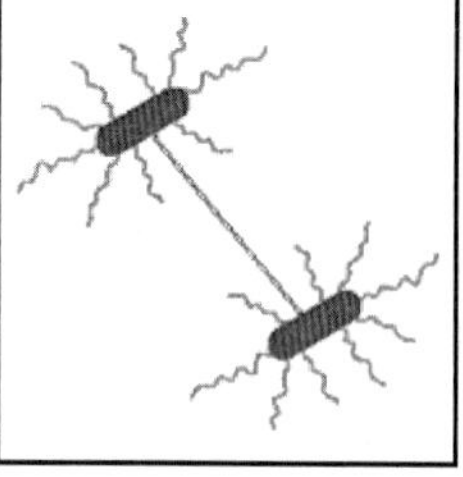

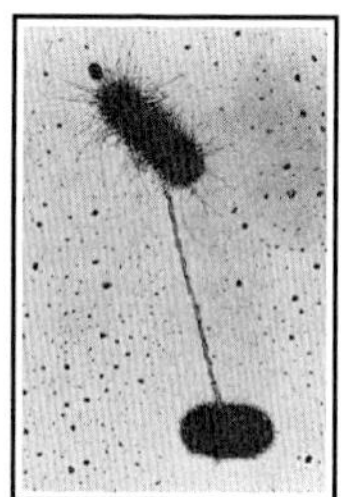

Pili

Sex pili

They also involve the virulence determination character of a bacterium. Ex. *Neisseria gonorrhoeae* causes disease if pili are present.

The pili also give resistance to bacterium from phagocytosis (Ex. *Streptococcus pyrogenes*)

The number of pili varies from few 1-4 to 100 - 200 per bacterium.

*3. Capsule*

Some bacterial cells are surrounded by viscous substance forming a covering layer or envelope around the cell wall, are called as capsules.

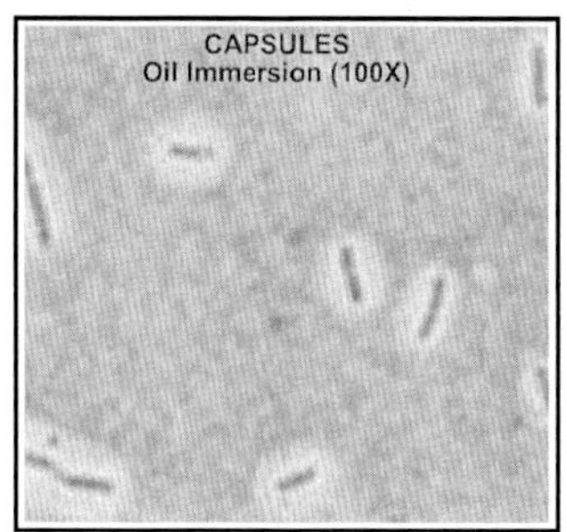


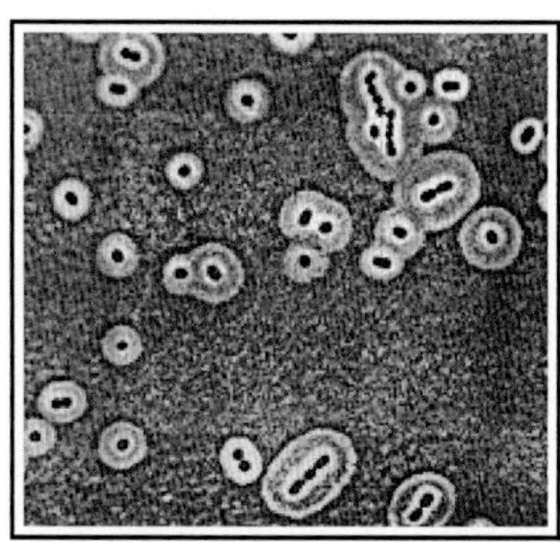

Capsule

If it is visible under light microscope using special staining, it is called as *capsule* and if it is so thin, not able to see under light microscope, called as *microcapsules*.

If it is so abundant that many cells are embedded in a common matrix, are called as *slime*.

*Note : The capsule or slime or microcapsule often referred as Glycocalyx*

Most of the capsules are polysaccharide nature, specifically homopolysaccharides. Ex. *Streptococcus mutans* capsule is made up of glucan.

Some bacterial capsule made up of polypeptides. Ex. *Bacillus anthracis* capsule composed of polymers of glutamic acid.

*Functions*

- Provide protection against temporary drying.
- Block the attachment of bacteriophages.
- Antiphagocytic (Ex. *Streptococcus pneumoniae*).
- Provide virulence to the bacteria (Ex. *S. pnemoniae* causes disease only if capsulated).
- Promote attachment of bacteria (ex. *Streptococcus mutans* adheres smooth surface of teeth and causes dental caries).

- Prevent cell aggregation in the suspension by electrical charges.
- Important role in biofilm application.

*Other appendages / structures outside the cell wall*

Sheath

Some species of bacteria, particularly those from fresh water and marine environment form chains or trichomes that are enclosed by hollow tube are called as Sheath. Ex. *Sphaerotilus.*

Prosthecae

Some species of bacteria have semirigid extension of cell wall and cell membrane are called as prosthecae. This is to increase the surface area of the cell for nutrient absorption. Ex. *Calulobacter.*

Stalk

The non-living ribbon or tubular appendage that are excreted by the bacterial cell are referred as stalk. Ex. *Gallionella.* They aided for attachment of the bacteria to surface.

### *4. Cell wall*

All the bacteria have rigid cell wall. The cell wall is the essential structure that protect the cell from mechanical damage and from osmotic rupture of lysis. Prokaryotes live relatively diluted atmosphere (has lower osmotic pressure than the osmotic pressure of inside the cell), which accumulate high salt concentration inside. The osmotic pressure against the inside of the plasma membrane may be the equivalent to 10 – 25 atm. Since the plasma membrane is delicate, plastic structure, it must be restrained by an outside wall made of porous, rigid material that has high tensile strength. The cell wall of bacteria deserves special attention for several reasons.

- They are essential for viability
- They are composed of unique compounds no where else in nature
- They are one of the important site for attack by antibiotics
- They provide adherence and receptor sites for drug or virus
- They cause symptoms of disease in animals
- They provide immunological distinction and variation among strains of bacteria

The bacteria can be divided in to two major groups based on their cell wall chemistry, called as gram positive and gram negative bacteria. The gram negative cell wall is a multilayered and quite complex, where as the gram +ve cell wall consists of a single type of molecule and often much thicker.

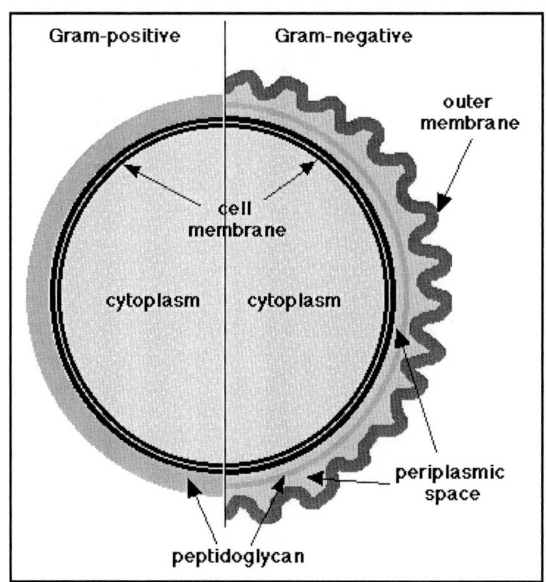


Cell wall difference between g+ve and g-ve bacteria

*Peptidoglycan layer*

The rigid layer of both gram +ve and –ve bacteria is similar in chemical composition, called as peptidoglycan (or murein). It consists of N-acetyl glucosamine and N-acetyl muramic acid and small group of amino acids such as L-alanine, D-alanine, D-glutamic acid, and lysine or diamino pimelic acid.

N-acetyl glucosamine and N-acetyl muramic acid were linked by β 1,4 glucoside linkage and the aminoacids were cross linked.

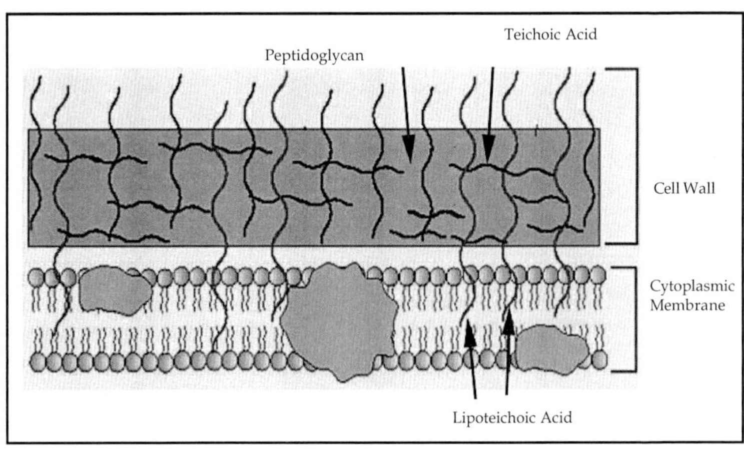


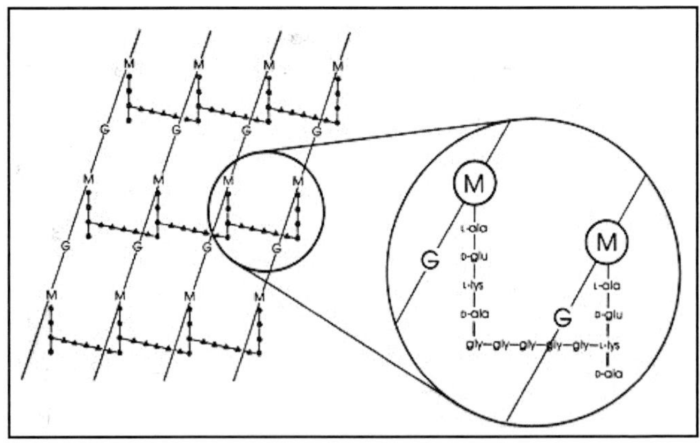


The structure of peptidoglycan

In G+ve bacteria, 90% of the cell wall consists of peptidoglycan layer and few quantities of teichoic acid usually present. Many bacteria have several layers (about 25) of peptidoglycon. In g-ve bacteria, only 10 per cent of wall is peptidoglycon and the major of the wall consists of outer membrane.

The peptidoglycan is signature of the bacteria can be destroyed by certain agents. Ex. Lysozyme, a protein that break the β-1,4 glucoside linkage of peptidoglycan and weaken the cell wall. Water then enters to the cell, and cell swells and eventually burst. Lysozyme is present in the tears, saliva and body fluids.

If proper concentration of solute that does not allow the water to penetrate into the cell, such as sucrose is added to the medium, the solute concentration will be balanced between in side and out side the cell. Under those condition, the lysozyme still digest the peptidoglycan, but lysis does not occur, instead, *protoplast* is formed. (Protoplast referred as the cell with out cell wall, but somatically protected). If these sucrose stabilized cells are placed in water, lysis occurs immediately.

*Spheroplast* often referred as the cell with some remaining of cell wall.

If peptidoglycan layer is removed in G+ve cell, protoplast will be formed and G-ve bacteria will form spheroplast.

*Outer membrane*

Besides peptidoglycan layer, G-ve bacteria contain additional wall layer made of lipo-polysaccharide. The layer is called as lipopolysaccharide layer or LPS layer or outer membrane.

The lipids and polysaccharides are intimately linked and formed this layer. The polysaccharide has two portions namely, o-polysaccharide and core polysaccharide.

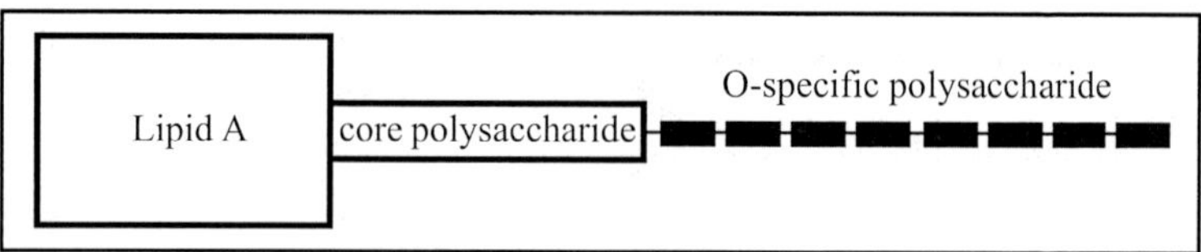


O-polysaccharide contains galactose, glucose, rhamnose, mannose, and one or more dideoxy sugar like abequose, colitose, paratose or trylose. These sugars were connected four or five membered sequence which often branched. The core polysaccharide consists of keto deoxyoctonate, glucose, galactose and N-acetyl glucosamine.

The lipid portion of the LPS referred as lipid A, is not a glycerol lipid but instead, the fatty acids are connected by ester amine linkage to disaccharides composed of N-acetyl glucosamine. The fatty acids are caproic, lauric, mysteric, palmitic, stearic acids.

A lipid protein is also found on the inner side of the outer membrane of gram negative bacteria. It acts like an anchor between outer membrane and peptidoglycan layer.

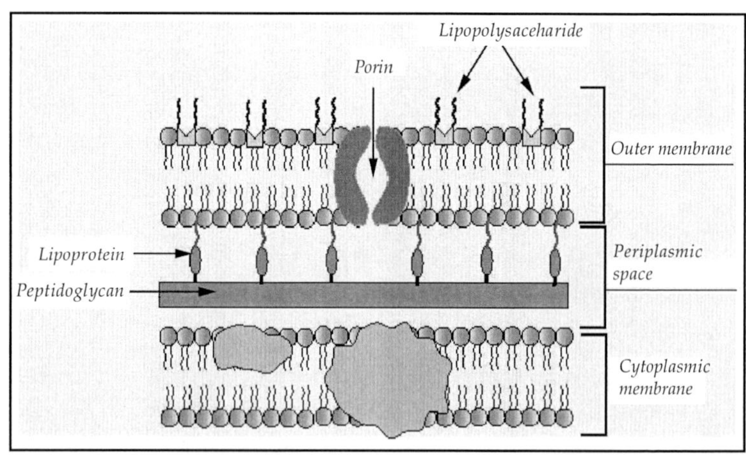


Structural arrangement of outer layer

*Endotoxins*

In *Salmonella, Shigella, Escherichia,* the outer membrane causes toxic and pathogenic symptom to human and mammals referred as endotoxins.

*Porins*

Protein substance present in the outer membrane of G-ve bacteria. They act like channel for entrance and exit of hydrophobic low molecular substances.

*Periplasm*

The space in between peptidoglycan and outer membrane is referred as periplasm or periplasmic space. In G+ve bacteria, it refers the space between peptidoglycan and cell membrane. This space harbours hydrolytic enzymes, binding proteins, detoxifying enzymes etc.

*5. Cell membrane*

The cell membrane is the most dynamic structure in prokaryotes. They act as selective permeability barrier that regulates the passage of substance in and out of the cell.

Bacterial membrane consists of 40 per cent phospholipids and 60 per cent protein. The phospholipids forms bilayer and the proteins are present scatterly (called integral protein) and some protein present on the surface of the layer (called peripheral protein).

The phospholipids are phosphoglycerides in which straight chain fatty acids are ester linked to glycerol.

Functions of cell membrane:

- Osmotic or permeability barrier
- Location of transport systems for specific solutes (nutrients and ions)
- Energy generating functions, involving respiratory and photosynthetic electron transport systems,

establishment of proton motive force, and transmembranous ATP-synthesizing ATPase.

- Synthesis of membrane lipids (including lipopolysaccharide in Gram-negative cells).
- Synthesis of murein (cell wall peptidoglycan).
- Assembly and secretion of extra cytoplasmic proteins.
- Coordination of DNA replication and segregation with septum formation and cell division.
- Chemotaxis (both motility and sensing functions).

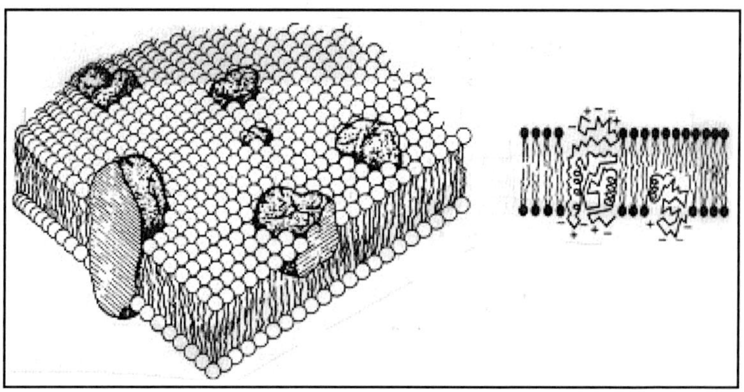

Fluid mosaic model of cytoplasmic membrane

*6. Mesosomes*

The cytoplamic membrane invagination in the form of tubular or vesicle shaped are referred as mesosomes. If the mesosomes are present in the centre of the cell referred as central mesosomes and they are involved in the cell division.

The mesosomes which are present in the peripheral region (aside) of the cell are referred as peripheral mesosomes and they involved in the transport of extracellular enzymes from cytoplasm to external regions.

### *7. Cytoplasm*

The cytoplasm of bacterial cell is the place where the functions for cell growth, metabolism, replication carried out. The gel like mat composed of water, protein, enzymes, nutrients, wastes, gases, like non-cellular materials and chromosomal DNA, ribosomes, plasmids like cellular components are present.

### *8. Chromosomal DNA*

The bacteria do not contain true membrane enclosed nucleus and chromosomes. They have a long coiled double strand single circular structure often called as chr. DNA or nucleoid or bacterial chromosome or chromatin body.

### *9. Plasmids*

A small circular, covalently closed, self replicating extra chromosomal DNA is also present in many bacteria. They are referred as *plasmids*. They are mostly controlling the special characters like pathogenicity, nodulation etc.

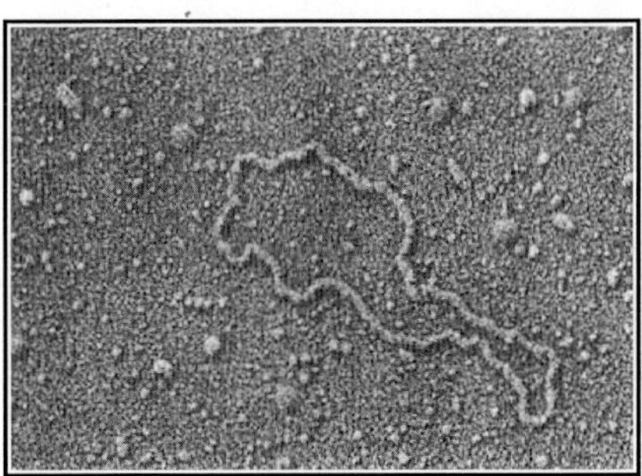

Electron micrograph of plasmid

*(The chromosomal DNA and plasmids are collectively called as genomic DNA).*

### *10. Ribosomes*

The ribosomes are the granular appearance in the cytoplasm involved in the protein synthesis. The size of prokaryotic ribosomes is of 70 S(sedimentation coefficient, Svedberg unit) and are of two sub units of 50S and 30S. The bacterial ribosomes never bound to any organelles in the cytoplasm.

*11. Cytoplasmic inclusions*

The cytoplasm of bacteria often contains inclusion granules referred as cytoplasmic inclusions. Inclusion bodies are usually storage materials of energy, carbon, phosphorus etc. Some aquatic bacteria possess large sized gas vacuoles which will help to float them.

The following table shows the cytoplasmic inclusions and their role in bacterial cell.

Some inclusions in bacterial cells

| Cytoplasmic inclusions | Where found | Composition | Function |
|---|---|---|---|
| Glycogen | many bacteria e.g. *E. coli* | polyglucose | reserve carbon and energy source |
| Polybetahydr-oxy butyric acid (PHB) | many bacteria e.g. *Pseudomonas* | polymerized hydroxy butyrate | reserve carbon and energy source |
| Polyphosphate (volutin granules) | many bacteria e.g. *Corynebacterium* | linear or cyclical polymers of $PO_4$ | reserve phosphate; possibly a reserve of high energy phosphate |
| Sulfur | phototrophic purple and green sulfur bacteria and lithotrophic colorless sulfur bacteria | elemental sulfur | reserve of electrons (reducing source) in phototrophs; reserve energy source in lithotrophs |

*Contd...*

| | | | |
|---|---|---|---|
| Gas vesicles | aquatic bacteria especially cyanobacteria | protein hulls or shells inflated with gases | buoyancy (floatation) in the vertical water column |
| Parasporal crystals | endospore-forming bacilli (genus *Bacillus*) | protein | toxic to certain insects |
| Magnetosomes | certain aquatic bacteria | magnetite (iron oxide) $Fe_3O_4$ | orienting and migrating along geo- magnetic field lines |
| Carboxysomes | autotrophic bacteria | enzymes for autotrophic $CO_2$ fixation | site of $CO_2$ fixation |
| Phycobilisomes | cyanobacteria | phycobili-proteins | light-harvesting pigments |
| Chlorosomes | Green bacteria | lipid and protein and bacteriochl-orophyll | light-harvesting pigments and antennae |

*Endospores*

Certain species of bacteria produce spores either within the cells are referred as endospores. They are unique to bacteria and *Bacillus, Clostridium* are common spore forming bacteria.

The endospores are extremely resistant to desiccation, staining, disinfecting chemicals, radiation and heat. The endospore can able to survive in boiling water, UV light and many harmful chemicals. They can survive even 100 years.

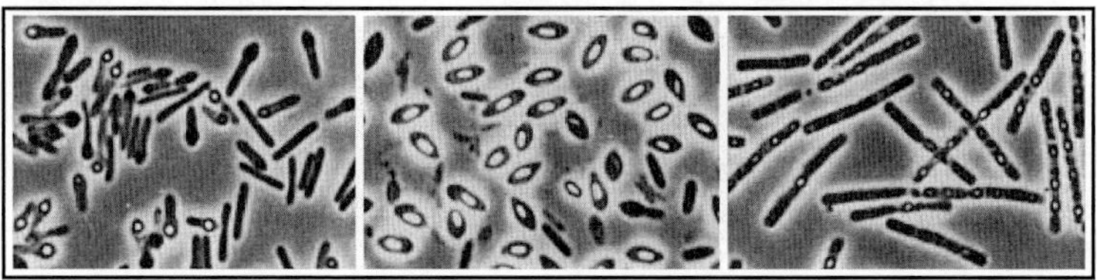

Bacterial endospores

First, the DNA fragmented and followed by invagination of cytoplasmic membrane. Then, the portion of the invagination is separated by complete formation of septum (this stage is called forespore). Then formation of cell wall around the forespore and followed by cortex, spore coat and exosporium synthesis occurred. Finally, cell lysed and the spore is released to the external environment.

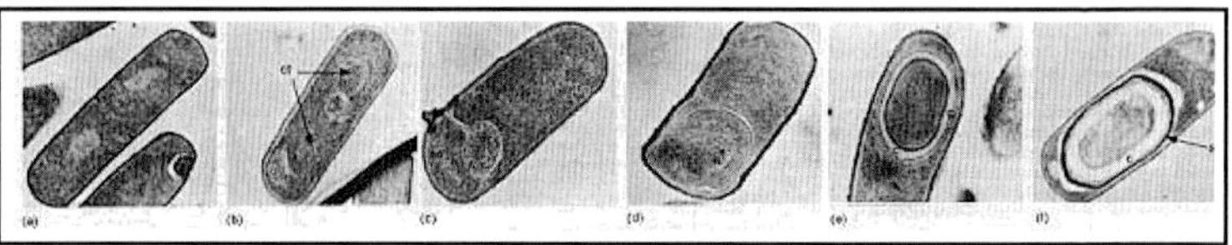

Process of endospore formation

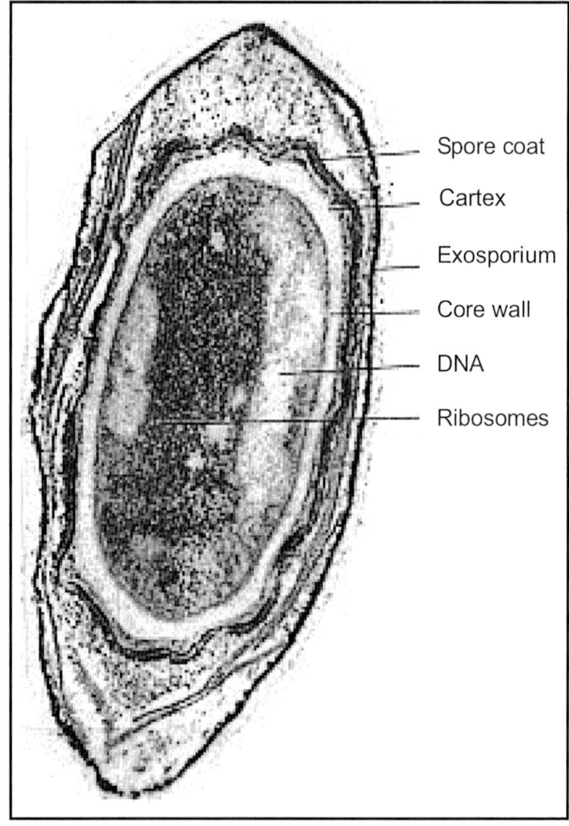

The outer most layer is called as *exosporium*, a thin delicate covering made of protein. The next layer is *spore coat*, composed of spore specific proteins. Below the spore coat is *cortex*, composed of cross linked peptidoglycan. Inside the cortex is spore protoplast which has cell wall (core wall), cell membrane, DNA, ribosome and little cytoplasm. One chemical substance that is characteristic of endospores not in vegetative cells is *dipicolinic acid* in the core region. They occupy 10% of total dry weight of the spores and responsible for heat resistance.

*Cyst* : Cyst is dormant, thick walled desiccation (drying) resistant form of cell. It can be differentiated from vegetative cells and can germinate under suitable condition. The cyst don't have temperature resistance. Ex. Cyst of *Azotobacter*

*Sporangiospores and Conidiospores* : the filamentous bacteria, actinomycetes produce these kind of spores. The spores were produced at the tip of the hyphae and if the spores are formed in a sac like structure (Sporangium), called as sporangiospores, if not called as conidiospores. These spores do not have heat resistance but can survive long period of drying.

## Bacteria - Reproduction

Since bacteria are single celled, a single cell division is enough to get reproduced into two daughter bacteria. The cell division of bacteria is referred as *transverse binary fission*. A single cell is divided into two by forming transverse septum (cross wall). The binary fission is an asexual reproduction.

The process of binary fission is as follows:

1. The cell became elongated from its original size and approximately twice length.
2. DNA replication takes place and DNA moves towards polar region.

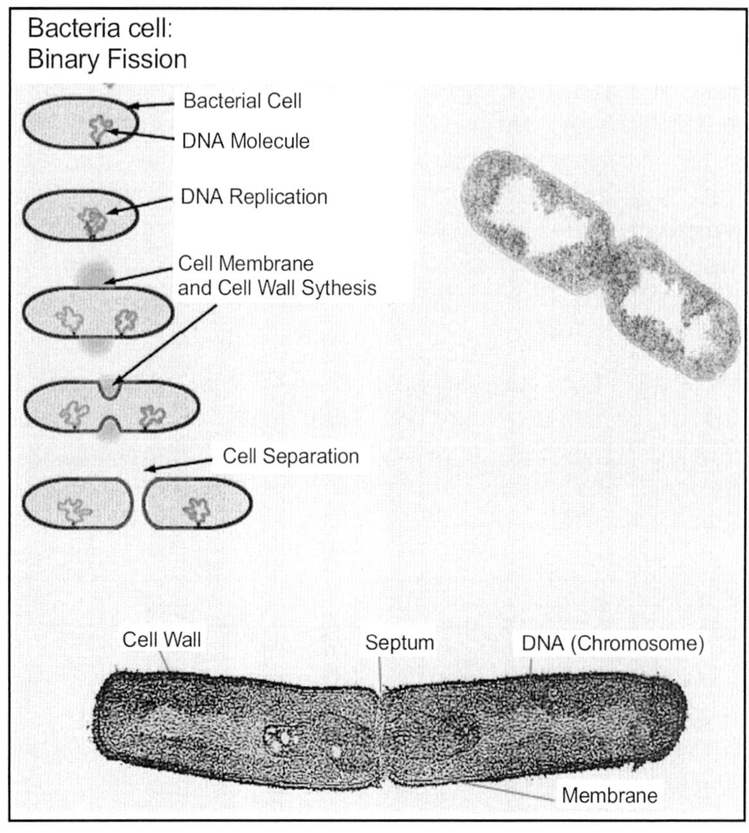


Binary fission process and cross section of bacterial cell before replication

3. Septum or cross wall (septum) occurs from the mesosome region
4. New cell wall formation occurs
5. After completion, the cells were separated out

Other methods of reproduction occur in bacteria:

*Budding* : A small bud developed at one polar region of the cell, elongated and finally developed to new cell. Ex. *Rhodopseudomonas acidophila.*

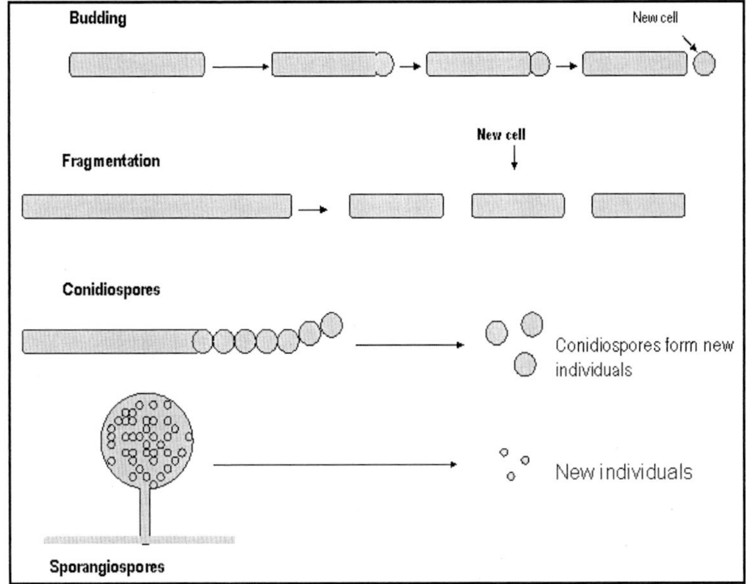


Ex. *Bacillus, E. coli, Streptococcus*

*Fragmentation :* Filamentous bacteria (actinomycetes) fragment into small rods and each rod gives raise to new cells. Ex. *Nocardia*

*Conidiospores and sporangiospores :* The filamentous bacteria produce spores in the hyphal tips and released to the environment, referred as conidiospores and sporangiospores. If the spores are produced in sac, referred as sporangiospores and without sac are referred as conidiospores.

## Growth

In microbiology, growth refers the increase in the number of cells. The bacterial cell is a synthetic machine that able to duplicate itself. More than 2000 chemical processes were occurred during one cell become two.

Step wise reactions occur during the growth of bacteria are as follows :

- The macromolecules such as DNA, RNA, protein, polysaccharides, lipids polymerization.
- Assembly of macromolecules.
- Cell organelles like cell wall, membrane, ribosome, etc. formation.

*Growth rate:* Increase of bacterial cell number per unit time is referred as growth rate

*Generation:* The interval of formation of daughter cells from parental cell is called as generation

*Generation time:* Time required to form two daughter cells from a single cell is called as generation time. This is also called as doubling time. *E. coli* has the doubling time of 20 min and *Rhizobium* has 2 h.

### *Growth curve of bacteria*

Assuming a single bacterium has been inoculated to a liquid medium and the bacterium starts multiplying by binary fission gives increase of population. If we plot a graph of $\log_{10}$ value of population against time, the graph will be of as follows:

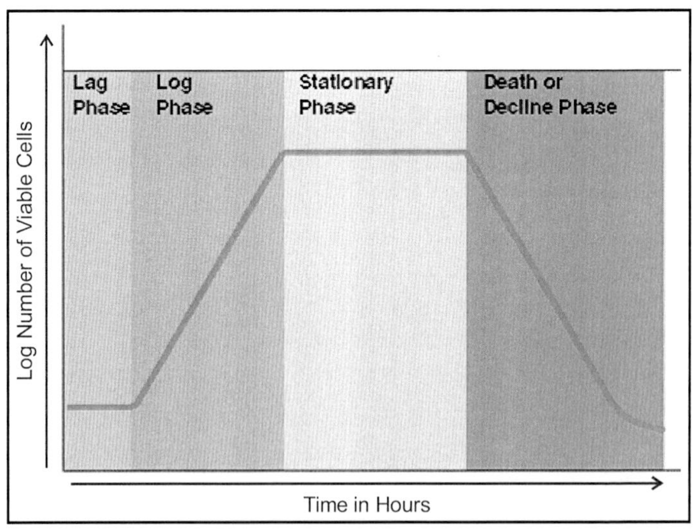

This curve is referred as growth curve of bacteria. The curve can be divided in to 4 phases. In the beginning, the growth (increasing the population) will not occur leads to straight line (called *lag phase*), then the cells start multiply gives linear line (*log or exponential phase*) followed by straight line (called *stationary phase*) and decline linear line (*death phase or decline phase*).

*Lag phase :* When a bacterium is introduced into a medium, it will not multiply immediately. After a period of time only the bacterium will start multiply. This period is called as lag phase.

- During this period, the cell increases its size.
- It will synthesis new protoplasm, enzymes, co enzymes etc.
- The bacterium will adjust the physiological environment.
- The cells are metabolically active.
- At the end of lag phase, each cell will start reproduction.

*Log / exponential phase :* After lag phase, the cell start multiply steadily by binary fission at constant rate. The portion of curve is linear and have constant generation or doubling time.

- All the cells are having uniform chemical composition, metabolic activity and physiological characters.
- At this time, the generation time will be of constant.
- The cells will be sensitive to chemical inhibitors.
- This is the ideal stage to study the metabolic activity of the bacterium.

*Stationary phase :* By this time, the essential nutrients of the medium get depleted and some toxic substances produced by the organism leads to slow down the cell division. At this time no net increase or decrease of cells in the medium occurs. This stage is called as stationary phase.

- During this stage, cell division rate and cell death rate is uniform.
- Death due to low nutrients content, pH change, toxic wastes, reduced oxygen occur.
- Cells are very small and having low number of ribosomes.
- Some cases, cells do not die and at the same time do not multiply.
- Some bacteria produce antibiotics at this stage as secondary metabolites for competitiveness.

*Death phase :* In this phase, few cells are alive and few died. The death rate is higher than the growth rate leads to decline the population. This phase is referred as death phase or decline phase.

- The cell lysis occur during this stage
- Some species can die very rapidly and some can persists for months

*Note :* These growth curve and the terms lag, log, stationary and decline phases are used only for the population or group of bacterial cells not for individual cells.

*Synchronous growth of bacteria :* All the cells of a bacterial culture in same stage of their growth cycle is referred as synchronous growth.

When bacterial growth which we discussed earlier as growth curve, if observed under microscope, will have different stages of its reproduction i.e., cell elongation, septum formation, divided cells etc. So, the cells and cell divisions will not be uniform. But, in synchronous growth culture, the bacterial cells will be of uniform stage.

The synchronous growth of bacteria will have the growth curve as following graph :

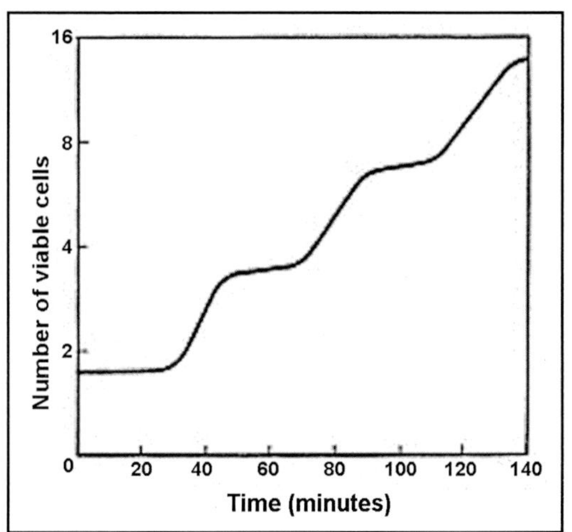


The synchronous growth of bacteria can be achieved by growing the cells at suboptimal temperature (below the optimum temperature). *E. coli* needs 37°C for its normal growth and if grown at 28°C, the synchronous growth will occur. This type of culture will be useful in molecular biology especially the gene cloning and transformation studies.

*Study state of growth of bacteria*

Maintaining the bacterial population in a particular stage / phase of the cycle is referred as study state of growth. The growth curve which discussed earlier is meant for the bacterial culture in a constant media referred as *batch culture*. In this type of culture, the maintenance of bacteria at particular stage is not possible.

For achieving the study state growth, the constant population should be maintained in the medium. This can be achieved by adding new medium and removing few cells at frequent time interval. This type of culture is referred as *continuous culture*.

The bacterium at late log phase is much interested for many industrial processes. So, maintaining the bacteria at log phase can be achieved by two systems called *chemostat* and *turbidostat*.

*Chemostat :* Maintaining the population of bacteria at particular stage using an important nutrient source as growth control is referred as chemostat. Ex. In a medium, if the glucose concentration is maintained by subsequent addition at particular dilution rate, so that it will be taken up by the organism immediately and at the same time a portion of the culture will be removed subsequently. This condition will maintain the bacterial growth at particular stage especially in log phase. This kind of devise is called chemostat.

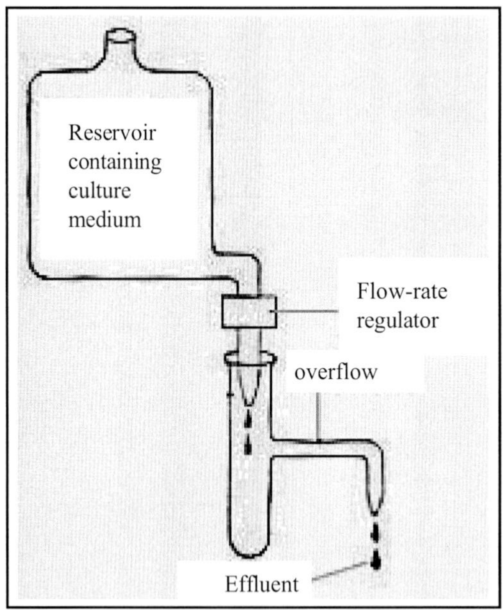


Chemostat - model

*Turbidostat* : An electrical devise will monitor the cell density (turbidity) and at appropriate turbidity the nutrient will be added ( as discussed above - glucose) at a dilution rate to maintain the constant population. If the cell density is higher, the instrument will add more quantity of solution to reduce or dilute the population and vice versa. This kind of device is referred as turbidostat.

Both the devices are highly useful in the continuous culture of bacteria.

## Nutrition and Physcial Conditions for Growth of Bacteria

All the bacteria essentially need energy, electron and carbon for their normal cellular activity, growth and reproduction.

*Energy* : Based on the source of energy, bacteria can be divided into two groups namely, chemotrophs and phototrophs.

a. Chemotrophs - utilize chemicals for their energy.

b. Phototrophs - utilize sun light for their energy.

*Electron* : Based on the source of electron, bacteria can be divided into two groups namely lithotrophs and organotrophs.

a. Lithotrophs - utilize inorganic substances as electron source.

b. Organotrophs - utilize organic substances as electron source.

*Carbon* : Based on the sole source of carbon for their utilization, the bacteria can be divided into two groups namely autotrophs and heterotrophs.

i. Autotrophs - use $CO_2$ as sole carbon source.

ii. Heterotrophs - use organic carbon as sole source.

- The combinations are as follows :
- Photolithotrophs;
- Chemolithotrophs;
- Photoorganotrops;
- Chemoorganotrophs;
- Photoautotrophs;
- Photoheterotrophs;
- Chemoautotrophs;
- Chemoheterotrophs

Some organisms exhibit different nutritional types like *Chemo lithotrophic heterotrophs*.

(Chemo - chemical for energy; litho - inorganic for electron donor and hetero - organic carbon)

Some organisms utilize inorganic compounds as energy and electron donors are inorganic compounds but utilize organic compounds for carbon source, referred as *mixotrophs*. Ex. *Desulfovibrio desulfuricans*.

*Physical conditions for growth*

The prokaryotes exist in nature under enormous range of physical condition such as $O_2$ concentration, hydrogen ion concentration (pH), temperature, salt concentration and water activity. The exclusive limit of life on the planet is set by the prokaryotes especially the archaea.

a. *Temperature* : Microorganisms have been found growing in virtually all environments where there is liquid water, regardless of its temperature. In 1966, Professor Thomas D. Brock at Indiana University made the amazing discovery in *boiling hot springs* of *Yellowstone National Park* that bacteria were not just surviving there, they were growing and flourishing. Boiling temperature could not inactivate any essential enzyme. Subsequently, prokaryotes

| Organism | Energy | | Electron | | Carbon | |
|---|---|---|---|---|---|---|
| | Photo | Chemo | Litho | Organo | Auto | hetero |
| *E. coli* | | X<br>Glucose | | X<br>Glucose | | X<br>Glucose |
| *Chromatium* (Purple sulphur bacteria) | X | | X<br>$H_2S$ | | X<br>$CO_2$ | |
| Cyanobacteria | X | | X<br>$H_2O$ | | X | |
| *Rhodospirillum* (Purple Non sulphur bacteria) | X | | | X<br>fatty acid | | X<br>fatty acid, alcohol |
| *Nitrosomonas* | | X | X<br>$NH_3$ | | X | |
| *Nitrobacter* | | X | X<br>$NO_2$ | | X | |
| *Thiobacillus* (Sulphur oxidizing bacteria) | | X<br>Sulfide, sulphur, thiosulfide | X | | X | |

X – refers the appropriate source

have been detected growing around black smokers and hydrothermal vents in the deep sea at temperatures at least as high as 115°C. Microorganisms have been found growing at very low temperatures as well. In super cooled solutions of $H_2O$ as low as -20°C, certain organisms can extract water for growth, and many forms of life flourish in the icy waters of the Antarctic, as well as household refrigerators, near 0°C.

Based on the growth of bacteria at different temperatures, there are three different temperatures are available.

1. *Minimum temperature :* The temperature below which the organism will not grow at all.
2. *Maximum temperature :* above which there won't be any growth of organism
3. *Optimum temperature :* The temperature in which maximum growth occurs.

The following table shows some of the organisms and their min, max and opt. temperature (°C) requirements.

| Bacterium | Minimum | Optimum | Maximum |
|---|---|---|---|
| *Listeria monocytogenes* | 1 | 30-37 | 45 |
| *Vibrio marinus* | 4 | 15 | 30 |
| *Pseudomonas maltophilia* | 4 | 35 | 41 |
| *Thiobacillus novellus* | 5 | 25-30 | 42 |
| *Staphylococcus aureus* | 10 | 30-37 | 45 |
| *Escherichia coli* | 10 | 37 | 45 |
| *Clostridium kluyveri* | 19 | 35 | 37 |
| *Bacillus flavothermus* | 30 | 60 | 72 |
| *Thermus aquaticus* | 40 | 70-72 | 79 |
| *Methanococcus jannaschii* | 60 | 85 | 90 |
| *Sulfolobus acidocaldarius* | 70 | 75-85 | 90 |
| *Pyrobacterium brockii* | 80 | 102-105 | 115 |

The effect of these temperatures on the growth rate of bacteria is represented as the following graph:

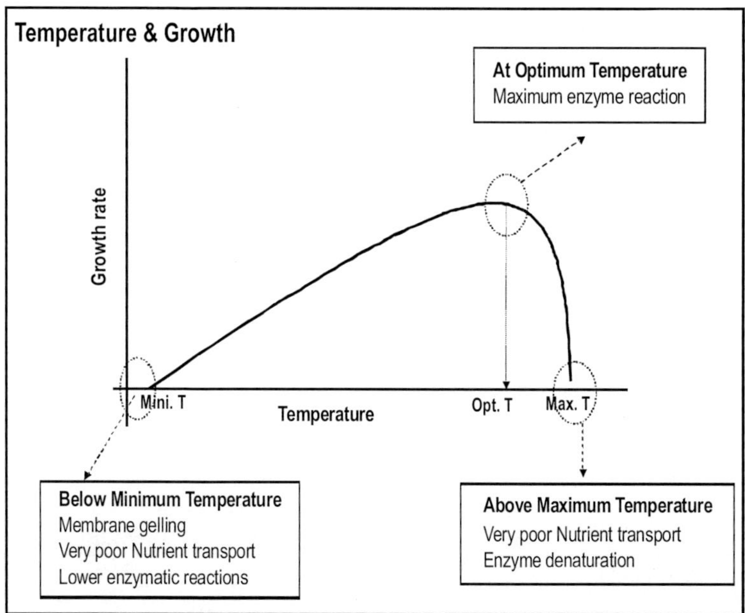


Based on the optimum temperature requirement, the organisms can be classified as :

1. *Psychrophiles :* The organisms which prefer low temperature (15°C) as optimum temperature are referred as psychrophiles. They have about 0°C and 20°C as their minimum and maximum temperature respectively. Ex. *Poloromonas vaculata*

2. *Psychrotolerant :* They are the psychrophiles but can grow even at higher temperature of 20-40°C (as the maximum temperature not optimum)

3. *Mesophiles :* The organisms which prefer 35°C (room temperature) as optimum temperature are referred as mesophiles. They have about 10°C and 48°C as their minimum and maximum temperature respectively. Ex. *Bacillus, Pseudomonas, E. coli*

4. *Thermotolerants :* They are the mesophiles but can grow even at higher temperatures (Ex. 45° to 55°C).

5. *Thermophiles :* The organisms which prefer high temperature (55-65°C) as optimum temperature are referred as thermophiles. They have 40°C and 70°C as their minimum and maximum temperatures respectively. Ex. *Bacillus stearothermophilus.*

6. *Hyper thermophiles :* The organisms which prefer very high temperatures (80 – 100°C) as their optimum temperature are referred as hyper thermophiles. They have 60°C and 110°C as their minimum and maximum temperature respectively. (Ex. *Thermococcus*).

Optimum growth temperature of some prokaryotes

| Genus and species | Optimal growth temp (°C) |
|---|---|
| *Vibrio cholerae* | 18-37 |
| *Rhizobium leguminosarum* | 20 |
| *Streptomyces griseus* | 25 |
| *Pseudomonas fluorescens* | 25-30 |
| *Erwinia amylovora* | 27-30 |
| *Staphylococcus aureus* | 30-37 |
| *Escherichia coli* | 37 |
| *Mycobacterium tuberculosis* | 37 |
| *Pseudomonas aeruginosa* | 37 |
| *Streptococcus pyogenes* | 37 |
| *Thermoplasma acidophilum* | 59 |
| *Thermus aquaticus* | 70 |
| *Bacillus caldolyticus* | 72 |
| *Pyrococcus furiosus* | 100 |

Hyperthermophilic Archaea

| Genus | Min. T (°C) | Opt. T (°C) | Maxi. T (°C) | Optimum pH |
|---|---|---|---|---|
| *Sulfolobus* | 55 | 75-85 | 87 | 2-3 |
| *Desulfurococcus* | 60 | 85 | 93 | 6 |
| *Methanothermus* | 60 | 83 | 88 | 6-7 |
| *Pyrodictium* | 82 | 105 | 113 | 6 |
| *Methanopyrus* | 85 | 100 | 110 | 7 |

b. *pH :* The pH, or hydrogen ion concentration, $[H^+]$, of natural environments varies from about 0.5 in the most acidic soils to about 10.5 in the most alkaline lakes. Most free-living prokaryotes can grow over a range of 3 pH units, about a thousand fold change in $[H^+]$. The range of pH over which an organism grows is defined by *three cardinal points*: the *minimum pH,* below which the organism cannot grow, the *maximum pH,* above which the organism cannot grow, and the *optimum pH,* at which the organism grows best. For most bacteria there is an orderly increase in growth rate between the minimum and the optimum and a corresponding orderly decrease in growth rate between the optimum and the maximum pH, reflecting the general effect of changing $[H^+]$ on the rates of enzymatic reaction.

Microorganisms which grow at an optimum pH well below neutrality (7.0) are called *acidophiles*. Those which grow best at neutral pH are called *neutrophiles* and those that grow best under alkaline conditions are called *alkaliphiles*. Obligate acidophiles, such as some *Thiobacillus* species, actually require a low pH for growth since their membranes dissolve and the cells lyse at neutrality. Several genera of Archaea, including *Sulfolobus* and *Thermoplasma,* are obligate acidophiles. Among eukaryotes, many fungi are acidophiles, and the champion of growth at low pH is the eukaryotic alga *Cyanidium* which can grow at a pH of 0.

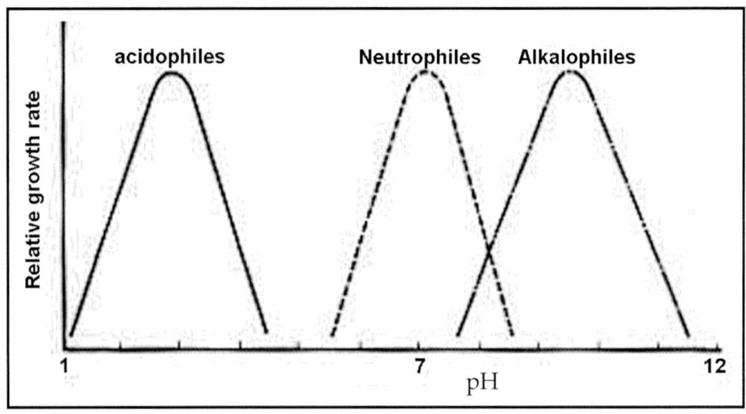


Relation between growth rate and pH

The following table shows the mini. pH, opt. pH, and maxi. pH requirement of few bacteria.

| Organism | Minimum pH | Optimum pH | Maximum pH |
|---|---|---|---|
| *Thiobacillus thiooxidans* | 0.5 | 2.0-2.8 | 4.0-6.0 |
| *Sulfolobus acidocaldarius* | 1.0 | 2.0-3.0 | 5.0 |
| *Bacillus acidocaldarius* | 2.0 | 4.0 | 6.0 |
| *Zymomonas lindneri* | 3.5 | 5.5-6.0 | 7.5 |
| *Lactobacillus acidophilus* | 4.0-4.6 | 5.8-6.6 | 6.8 |
| *Staphylococcus aureus* | 4.2 | 7.0-7.5 | 9.3 |
| *Escherichia coli* | 4.4 | 6.0-7.0 | 9.0 |
| *Clostridium sporogenes* | 5.0-5.8 | 6.0-7.6 | 8.5-9.0 |
| *Erwinia caratovora* | 5.6 | 7.1 | 9.3 |
| *Pseudomonas aeruginosa* | 5.6 | 6.6-7.0 | 8.0 |
| *Thiobacillus novellus* | 5.7 | 7.0 | 9.0 |
| *Streptococcus pneumoniae* | 6.5 | 7.8 | 8.3 |
| *Nitrobacter* sp | 6.6 | 7.6-8.6 | 10.0 |

c. *Oxygen* : Oxygen is a universal component of cells and is always provided in large amounts by $H_2O$. However, prokaryotes display a wide range of responses to molecular oxygen $O_2$.

Molecular oxygen is both beneficial and harmful for biologicals.

*Beneficial :* It act as terminal electron acceptor in the respiratory chain – Electron transport chain reaction enable to form ATP.

*Harmful :* The oxygen derivatives such as hydrogen *peroxide, singlet oxygen and super oxide* are highly toxic and oxidizing agents.

The organisms can be divided into 4 or 5 different groups based on their oxygen requirement.

*Obligate aerobes* require $O_2$ for growth; they use $O_2$ as a final electron acceptor in aerobic respiration.

Ex. *Micrococcus*

*Obligate anaerobes :* do not need or use $O_2$ as a nutrient. In fact, $O_2$ is a toxic substance, which either kills or inhibits their growth. Obligate anaerobic prokaryotes may live by fermentation, anaerobic respiration, bacterial photosynthesis, or the novel process of methanogenesis. Ex. *Clostridium, Methanobcaterium.*

*Facultative anaerobes (or facultative aerobes)* are organisms that can switch between aerobic and anaerobic types of metabolism. Under anaerobic conditions (no $O_2$) they grow by fermentation or anaerobic respiration, but in the presence of $O_2$ they switch to aerobic respiration. Ex. *E. coli.*

*Aerotolerant anaerobes* are bacteria with an exclusively anaerobic (fermentative) type of metabolism but they are insensitive to the presence of $O_2$. They live by fermentation alone whether or not $O_2$ is present in their environment. Ex. *Streptococcus.*

*Microaerophiles* : They need very low level of oxygen for their growth, but cannot tolerate the level of atmospheric $O_2$ level. Ex. *Spirillum volutans.*

Table 6. Terms used to describe $O_2$ Relations of Microorganisms

| | Environment | | |
|---|---|---|---|
| Group | Aerobic | Anaerobic | $O_2$ Effect |
| Obligate Aerobe | Growth | No growth | Required (utilized for aerobic respiration) |
| Microaerophile | Growth if level not too high | No growth | Required but at levels below 0.2 atm |
| Obligate Anaerobe | No growth | Growth | Toxic |
| Facultative Anaerobe (Facultative Aerobe) | Growth | Growth | Not required for growth but utilized when available |
| Aerotolerant Anaerobe | Growth | Growth | Not required and not utilized |

*Oxygen detoxification* : Most of the aerobes (and aerotolerant organisms too) have three major enzymes either all or two to detoxify the oxygen derivatives. They are *catalase, peroxidase and super oxide dismutase.* Obligate anaerobes lack these enzymes, and therefore undergo lethal oxidations by various oxygen radicals when they are exposed to $O_2$.

d. *Water activity* : Water is the solvent in which the molecules of life are dissolved, and the availability of water is therefore a critical factor that affects the growth of all cells. The availability of water for a cell depends upon its presence in the atmosphere (relative humidity) or its presence in solution or a substance (*water activity*). The water activity (Aw) of pure $H_2O$ is 1.0 (100% water). Water activity is affected by the presence of solutes such as salts or sugars, that are dissolved in the water. The higher the solute concentration of a substance, the lower is the water activity and vice-versa. Microorganisms live over a range of Aw from 1.0 to 0.7. The Aw of human blood is 0.99; seawater = 0.98; maple syrup = 0.90; Great Salt Lake = 0.75. Water activities in agricultural soils range between 0.9 and 1.0.

The water requirement (Aw) of few bacteria are as follows:

*Caulobacter* - 1.00; *Pseudomonas, Salmonella, E. coli* - 0.91; *Lactobacillus* - 0.90; *Bacillus* - 0.90; *Staphylococcus* - 0.85; *Halococcus* - 0.75.

*Note:* The concept of lowering water activity in order to prevent bacterial growth is the basis for preservation of foods by drying (in sunlight or by evaporation) or by addition of high concentrations of salt or sugar.

e. *salt concentrations* : The only common solute in nature that occurs over a wide concentration range is salt [NaCl], and microorganisms are named based on their growth response to salt. Microorganisms that require some NaCl for growth are *halophiles.*

*Mild halophiles* require 1-6% salt, (Ex: *Staphylococcus areus*).

*Moderate halophiles* require 6-15% salt; (Ex. *Vibrio fisheri*).

*Extreme halophiles* require 15-30% NaCl for growth. (Ex. *Halobacterium salanarum*).

Bacteria that are able to grow at moderate salt concentrations, even though they grow best in the absence of NaCl, are called *halotolerant*. Although halophiles are "osmophiles" (and halotolerant organisms are "osmotolerant") the term *osmophiles* is usually reserved for organisms that are able to live in environments high in sugar.

Organisms which live in dry environments (made dry by lack of water) are called *xerophiles.*

## Archaea

(Sing: archaeon)

Archaea are the diversified group of prokaryotes primitive to bacteria.

The phylogenetic tree of life consists of three domains (the highest unit of taxonomy) namely, a. archaea b. bacteria c. eukaryota. Among the three, archaea and bacteria are prokaryotes and eukaryota consists of all eukaryotic organisms including plant and animals.

The archaea decide the *limit of life on the earth.* These organisms survive on the maximum extreme conditions occur on the earth. Ex. High temperature hot springs; high acidic soils; high salt lakes, extreme anaerobic conditions etc.

Phenotypic properties of Bacteria and Archaea compared with Eukarya

| Property | Cellular Domain Eukarya | Bacteria | Archaea |
|---|---|---|---|
| Cell configuration | eukaryotic | prokaryotic | prokaryotic |
| Nuclear membrane | present | absent | absent |
| Number of chromosomes >1 | | 1 | 1 |
| Chromosome topology | linear | circular | circular |
| Murein in cell wall | - | + | - |
| Cell membrane lipids | ester-linked glycerides; unbranched; polyunsaturated | ester-linked glycerides; unbranched; saturated or monounsatu rated | ether-linked branched; saturated |
| Cell membrane sterols | present | absent | absent |
| Organelles (mitochondria and chloroplasts) | present | absent | absent |

*contd...*

*Table contd...*

| | | | |
|---|---|---|---|
| Ribosome size | 80S (cytoplasmic) | 70S | 70S |
| Cytoplasmic streaming | + | - | - |
| Meiosis and mitosis | present | absent | absent |
| Transcription and translation coupled | - | + | + |
| Amino acid initiating protein synthesis | methionine | N-formyl methionine | methionine |
| Protein synthesis inhibited by streptomycin and chloramphenicol | - | + | - |
| Protein synthesis inhibited by diphtheria toxin | + | - | + |

**Morphological similarities & variations between archaea and bacteria**

When observed under microscopes, archaea and bacteria look similar but few morphological similarities and variations occur between archaea and bacteria.

## Similarities

- The archaeal cell size, shape and structures are similar to bacteria.
- Both contain single circular DNA as chromosomal material.
- Both contain 70s ribosomes.
- Cell wall and cell membranes are present in both the cases.

## Variations

- The cell wall of archaea is made up of proteins, glycoproteins and polysaccharides. Few archaea have peptidoglycan layer but with modifications. In bacteria, the N-acetyl glucosamine and N-acetyl muramic acid are linked by means of ß 1-4 linkage; whereas in archaea, instead of NAM, N acetyl talosaminuronic acid linked with N acetyl glucosamine with ß 1-3 linkage. This type of cell wall is referred as *pseudomurein* or *pseudopeptidoglycan* layer.

- Since, the cell wall composition is different, the lysozyme can not act on the cell wall, whereas the bacteria can be lysed by lysozyme.
- The bacterial cell membrane is made up of long chain fatty acid which has ester linkage with glycerol. Whereas the archeal cell membrane is long chain alcohols with ether linkage with glycerol
- *Methionine* is the first amino acid in archeal protein synthesis whereas the bacteria uses *N-formyl methionine* as its first aminoacid formation.
- *Histone* is present in the chromosome material of archaea, whereas the bacteria do not have histone protein.

**Diversity**

Like bacteria, the archaea also has variety of metabolism. They have chemoorganotrophy, chemolithotrophy type of nutrition. Based on the habitat and phylogenic relationship, the archaea are divided into four groups as

1. Extreme halophilic archaea
2. Methanogens
3. Hyper thermophiles
4. Thermoplasma

1. *Extreme halophilic archaea*

These archaea survive in extremely higher salt concentrations like Great salt lake (USA). The salt concentration is about 40.1 g/l of $Na^+$ and 181 g/l of Cl. These archaea has the optimum salt concentration of 1.5 M to 5.5 M NaCl. These archaea are collectively called as *halobacteria*.

The genera are *Halobacterium, Halorubrum, Halococcus, Natronobacterium, Natronosomonas.*

- These bacteria are heterotrophs.
- They have aerobic respiration with electron transport chain.
- All the organelles were stabilized by $K^+$.
- They provide purple membrane in the cell membrane, which is used for ATP synthesis.
- *Bacteriorhodopsin* is the light harvesting pigment present in the membrane gives purple colour. These pigment directly produce the proton gradient in the membrane (as that of electron transport chain) but without any electron carriers finally leads to ATP synthesis. This process is known as *non-photosynthetic phosphorylation*.

Protection against salt:

- Internal accumulation of K ions
- Acidic aminoacids of cell membrane
- *K stabilized cell organelles* : are some of the mechanisms by which these archaea protected themselves from salt.

*2. Methanogens*

They are the obligate anaerobic thermophiles present in fresh water sediments, deep soil, ruman of cattle, paddy fields. (Cattle can produce 50 l of methane per day).

- They are the obligate anaerobes.
- They use $H_2$ and $CO_2$ as electron and carbon sources for energy generation and anabolism through acetyl CoA pathway.

- Some organisms use CO, $CH_3OH$, HCHO as other substrates for C assimilation.
- They use $CO_2$ as terminal electron acceptor for methane formation.
- The role of methanogens in decomposition of organic waste is highly useful in environment and at the same time methane emission from flooded soil is highly harmful to ozone layer of atmosphere.

*Methanobacterium, Methanobrevibacterium, Methanococcus, Methanospirillum, Methanosarcina* are some of the important genera present in this group.

3. *Hyperthermophiles*

They are extreme thermophiles present in hot sulphur rich environments such as hot springs, volcanisms etc.

- They require 80 – 100° C as optimum temperature for growth.
- Unusually stable at high temperatures.
- They need sulphur as essential nutrients.
- Both aerobes and anaerobes are present.
- Chemolithotrophy and chemoorganotrophy type nutrition.
- Chemolithotrophs reduce the sulphur as terminal electron acceptor, form $H_2S$ and derive energy. Ex. *Pyrococcus, Ferroglobus, Archeoglobus.*
- Chemoorganotrophs use organic compounds and sulphur for energy and carbon source. Ex. *Thermoproteus, Thermococcus, Desulfurococcus, Pyrococcus, Sulfolobus.*

*Sulfolobus* was the first hyperthermophile isolated by T.D. Brock from hot spring of Yellowstone National Park.

*Thermus aquaticus* is an archaean from which the DNA polymerase enzyme isolated, is very much useful in the *DNA synthesis in lab condition* called as *Polymerase chain reaction*.

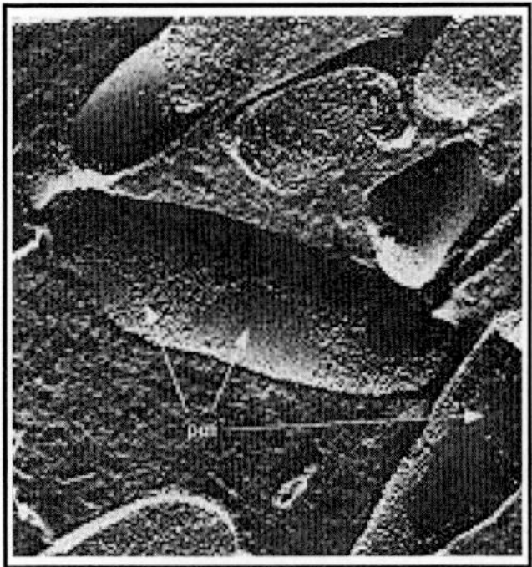

*Halobacterium* - bacteriorhodopsin embedded membrane

## 4. *Thermoplasma & Picrophilus*

They are thermophilic acidophiles. They prefer 55°C as optimum temperature and pH 2.0 as optimum condition for growth. They are present in the coal mine spoils (waste created during coal mining).

The another feature of these archaea is *lack of cell wall.* They have only membrane and the membrane itself protect the cell from high temperature and acidity.

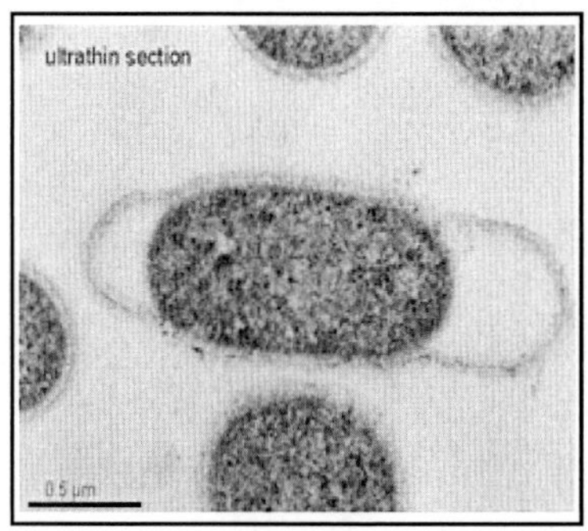


*Thermotoga*

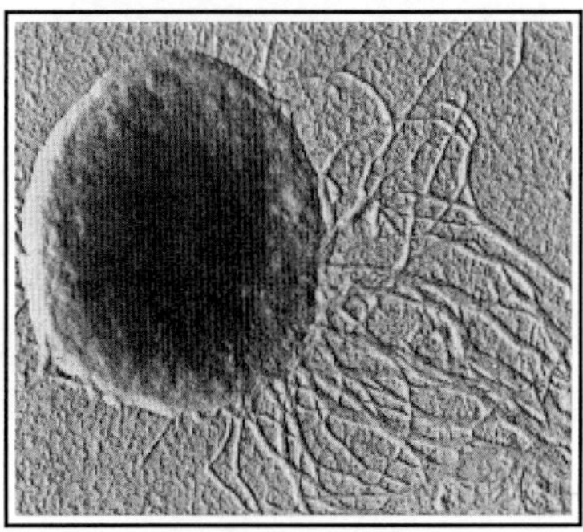

*Methanococcus*

Yellow National Park – Hot spring

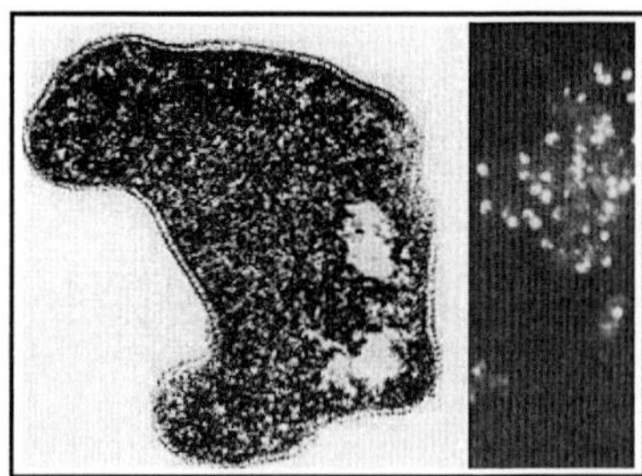

*Sulfolobus*

## Bacterial Taxonomy

Taxonomy refers the science of classification, which includes identification, nomenclature and systematic arrangement of any group of organisms.

Identification – Identify the individual

Nomenclature – naming the individual

Classification – arranging the individual by groups

*Conventional classification method*

For identification, nomenclature and classification of any bacterium, conventionally few characters are used. The characters are as follows:

1. *Morphological characters* : Cell shapes, size and structures; cell arrangement; occurrence of special structures; developmental forms; staining reactions; motility and flagellar arrangement.
2. *Chemical characters* : Various chemical constituents of cell like cell wall chemicals.
3. *Cultural characters* : Nutritional requirement and physical conditions required for growth and the manner by which growth of bacteria occur in solid and liquid medium.
4. *Metabolic characters* : The way by which the bacterium obtain their energy, carbon and electron source; regulation of these reactions.

5. *Antigenic characters* : Special large chemical components (antigens) of the bacterium - distinctive for certain kind of microorganisms.
6. *Genetic characters* : Characteristics of hereditary material of the cell (DNA) and occurrence and functions of other kind of DNA that may be present such as plasmid.
7. *Pathogenicity* : The ability to cause the disease in various plant and animal or even other microorganisms.
8. *Ecological characters* : Habitat and distribution of the organism in nature and the interaction between and among species in natural environment.

Based on these above characters, the conventional taxonomy of bacteria developed.

*Numerical Taxonomy* : A method used in taxonomy to determine and numerically express the degree of similarity of every strain of prokaryotes is referred as numerical taxonomy.

$$\%\,\text{similarity} = \frac{\text{No. of characters similar}}{\text{No. of character similar} + \text{No. of characters not similar}}$$

*Recent method of bacterial classification*

To understand the present concept of classification, the knowledge of evolution is essential. The present bacterial classification is mainly based on the *evolutionary relationships* of the bacteria.

A brief about evolution :

---

- 4.6 billion years ago, the earth was originated
- 3.86 billion years ago, water formation occur
- 3.5 billion years ago rocks formed
- Several shells of earth were formed. 4 layers . 1. ***Atmosphere*** – $H_20$, Nitrogen, $CO_2$, $SO_4$, $CH_4$ 2. ***Crust*** - U, K, O, Na, TH; 3. ***Mantle*** - Si, AL, O, Mg, Fe; 4. ***Core*** : Fe, Si, Ni.)

- At that time, (3.5 b years), in the presence of lightening & UV lights, methane, ammonia, hydrogen, water were converted to biochemicals like amino acid, formaldehydes.
- Then formed RNA, able to replicate them selves.
- Then surrounded by lipoprotein for the new life. (RNA world)
- Then DNA formed and RNA acts as a coding agent. Protein played major role and thus a new life formed.
- The atmosphere of that period was reduced one.
- The fossils are the evidence for microbial life of that period. Ancient rocks contain recognizable microfossils that appear as bacterial shapes. They are referred as *stromatolites*. These fossils were recognized as *filamentous phototrophic bacteria* (Not as present cyanobacteria - because the present one has oxygenic photophosphorylation whereas at that time the atmosphere was devoid of oxygen). The single celled phototrophs formed about 3.5 b yr ago.
- About 1 billion years agro, eukaryotes formed.
- 600 million years ago variety of plants and animals formed, evolved and disappeared.
- Mammals formed by 100 million years ago and 2-3 million years ago, MAN appeared.

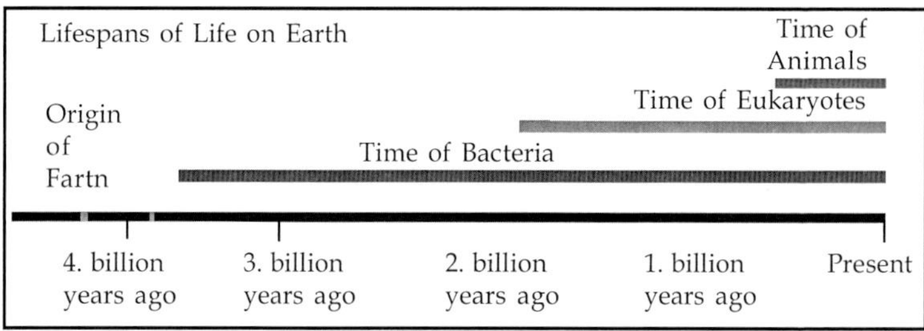


Recent method of classification of prokaryotes is mainly based on the evolutionary relationship. There are three methods presently available for grouping the microorganisms.

- DNA: DNA hybridization
- Ribotyping (16s RNA analysis)
- FAME (Fatty acid methyl ester analysis)

Among the above three, ribotyping is most common and frequently used tool to classify the microorganisms.

1. *DNA:DNA hybridization :* The DNA of one organism is subjected to hybridize with other organism and the degree of hybridization will vary with organisms depends upon their relativity. This variation will be used to identify and group the organism.
2. *FAME:* The fatty acid composition of prokaryotes give very high diversity. The fatty acid compositions especially the cell wall fatty acids analysis is used to identify the organisms.
3. *Ribosomal analysis :* Among the cellular organelles, ribosome is present in all the living organisms; ancient molecule; functionally constant; universally distributed and well conserved.

Ribosomal RNAs in Prokaryotes

| Name | Size (nucleotides) | Location |
|---|---|---|
| 5S | 120 | Large subunit of ribosome |
| 16S | 1500 | Small subunit of ribosome |
| 23S | 2900 | Large subunit of ribosome |

The ribosome can be divided into two subunits namely 30S and 70S (In eukaryotes 80S). The 30S subunit is referred as Smaller Sub-Unit (SSU). Further, the 30S can be divided in to 3 units namely, 5S, 16S and 23S. (based on their sedimentation co-efficient). (Note: For eukaryotes, 18S instead of 16S).

By analysing the genetic information of 16S RNA, the *phylogeny* of a prokaryote is possible. *Phylogeny refers the evolutionary history of the organism.*

The 16S RNA genetic analysis can be done by following procedure:

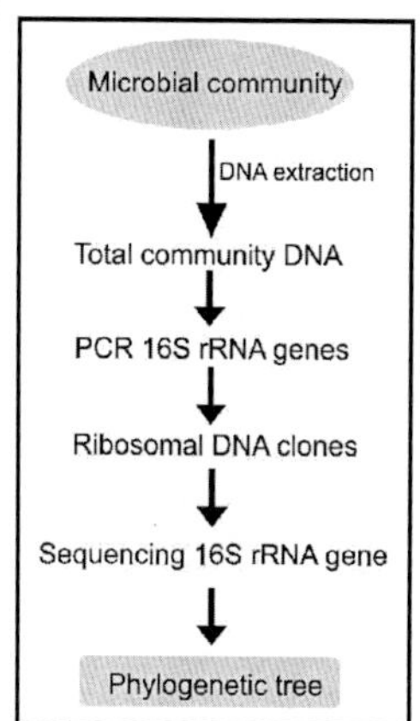

After getting the gene sequence information, the DNA base informations will be compared with already analysed data base. Based on the similarity percentage, the *evolutionary distance* of an organism will be calculated. Based on the evolutionary distance of various organisms, a branched tree like map can be obtained, which is referred as *Phylogenetic tree.*

Diversity of rRNA genes of different organisms

| | |
|---|---|
| Human | ...GTGCCAGCAGCCGCGGTAATTCCAGCTCCAAT AGCGTATATTAAAGTTGCTGCAGTTAAAAAG... |
| Yeast | ...GTGCCAGCAGCCGCGGTAATTCCAGCTCCAAT AGCGTATATTAAAGTTGTTGCAGTTAAAAAG... |
| Corn | ...GTGCCAGCAGCCGCGGTAATTCCAGCTCCAAT AGCGTATATTTAAGTTGTTGCAGTTAAAAAG... |
| *Escherichia coli* | ...GTGCCAGCAGCCGCGGTAATACGGAGGGTGCA AGCGTTAATCGGAATTACTGGGCGTAAAGCG... |
| *Anacystis nidulans* | ...GTGCCAGCAGCCGCGGTAATACGGAGAGGCA AGCGTTATCCGGAATTATTGGGCGTAAAGCG... |
| *Thermotoga maratima* | ...GTGCCAGCAGCCGCGGTAATACGTAGGGGGCA AGCGTTACCCGGATTTACTGGGCGTAAAGGG... |
| *Methanococcus vannielii* | ...GTGCCAGCAGCCGCGGTAATACCGACGGCCCG AGTGGTAGCCACTCTTATTGGGCCTAAAGCG... |

The 16S rRNA sequence that is *unique* to individual organism at strain level is often referred as *signature sequence.* By comparing the genetic informations of 16S RNA, it is possible to identify and classify the organisms.

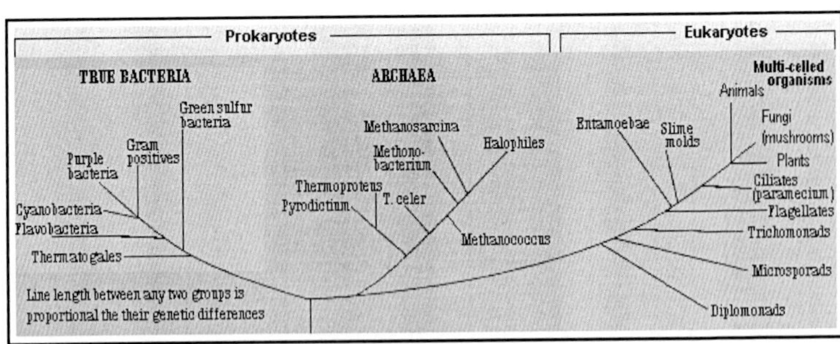


Phylogenetic tree based on the evolutionary distance:

## Three domains of life

Based on the signature sequence, the phylogenic tree divided into three domains namely, bacteria, archaea and eukaryota. Among the three, two of which are prokaryotes (bacteria and archaea). This universal phylogenic tree is sharing a common ancestor called *universal ancestor*, which is at present not known. The microorganisms fall in all the three domains.

## Domain : Eucarya (Eukaryotes)

*Major Divisions of Eucarya*

The Eucarya are usually classified into four kingdoms: plants, animals, fungi and protists. The first three of these correspond to phylogenetically coherent groups as well. However, the eukaryotic protists do not form a group, but rather are comprised of many phylogenetically disparate groups (including slime molds, multiple groups of algae, and many distinct groups of protozoa). Just as molecular analyses where required to see the natural relationships among prokaryotes, they are also allowing us to infer relationships among these non-plant, non-animal, non-fungus eukaryotes.

Origin of eukaryotes from prokaryotes was mainly explained by *endosymbiont theory*. The chloroplast and microchondria of eukaryotes contain smaller amount of DNA. These 16S rRNA sequence analysis showed that mytochondria and proteobacteria like *Agrobacterium* and *Rhizobium* are close relatives. Similarly, the chloroplast and photosynthetic bacteria, *Rhodopseudomonas* and *Rhodobacter* are phylogenitically highly related. This theory was evidenced by 16S RNA sequence and phylogenic analysis.

## Domain: Archaea

*Major Groups within the Archaea* (Note: Please refer the notes on archaea for details).

The major physiological groups within the Archaea (informally, archaes) are extreme thermophiles (including thermoacidophiles), sulfate reducers, methanogens, and extreme halophiles. These physiological groups do not correspond to the phylogenetic groups. This is largely because the methanogens have given rise to multiple other groups, including the halophiles, the sulfate reducers, and at least two other line ages of nonmethanogenic thermophiles.

The thermophilic members of the Archaea include the most thermophilic organisms cultivated in the laboratory, some having optimal growth at >100 ÉC. The aerobic thermophiles are also acidophilic; they oxidize sulfur in their environment to sulfuric acid, creating acid hot springs. The hot acidic waters can dissolve limestone and some other minerals out of the surrounding rock matrix, creating boiling mud pots of the insoluble residue from the rock.

The extreme halophiles include the most salt tolerant organisms known. They are aerobic or microaerophilic. They have also invented a form of photosynthesis that does not use chlorophyll.

The sulfate-reducing Archaea are now known to be a wide spread group. One of the reasons that this group has received attention is that sulfate-reducing organisms are responsible for "souring" oil wells (increasing their sulfur content). In addition to increasing the sulfur emissions when the oil is burned (or increasing the cost of refining it), sulfide attacks the metal in well casings and pipelines.

Methanogens are strict anaerobes, yet they gave rise to at least two separate aerobic groups: the halophiles and a thermoacidophilic lineage. Methanogens produce most of the methane in the Earth's atmosphere. Methane is a significant green-house gas (i.e., it is a cause of global warming). Much of this methane production is associated with activities for which humans are at least partially responsible. This includes the flooding of rice paddies and the raising of domestic cattle.

## Domain: Bacteria

This domain contains the prokaryotic organisms, bacteria. They are diversified group varies in morphology, metabolism, nutritional and physiological conditions. The major groups of bacteria will be discussed in detail in the later part of this chapter.

### *Nomenclature*

Following the binomial system of nomenclature, all the bacteria are given genus and species names. The names are either Latin or Greek derivations of some descriptive property appropriate for the species. Ex. *Bacillus subtilis, B. cereus, B. sterothermophilus* and *B. acidocaldarius* means "Slender", "waxan", "heat loving" and "acid thermal" respectively.

### *Taxonomical groups*

Linnaeus defined the biological classification system that we still use for plants and animals, and, with relatively minor modifications, for fungi and other microorganisms. It is a hierarchical system that starts with a few categories at the highest level, and further subdivides them at each lower level. The levels in the hierarchy are as follows:

Domain
Kingdom
Phylum
Class
Order
Family
Genus
Species

*Bergey's Manual of systematic bacteriology*

Bergey's manual of systematic bacteriology is a compendium of standard and molecular informations about the available prokaryotes. The Bergey's manual of systematic bacteriology - First edition consists of 4 volumes:

Vol 1. Gram positive bacteria

Vol 2. Gram negative bacteria

Vol 3. Bacteria with unusual properties including archaea

Vol 4. Filamentous bacteria

This system does not have much phylogenic informations. The second edition of the Bergey's manual provides much clear informations about the genetic relationship among the organisms. The second edition was divided into 5 volumes.

Vol 1. Archaea, cyanobacteria, phototrophs and deeply branched genera

Vol 2. Proteobacteria

Vol 3. Low G+C gram positives

Vol 4. High G+C gram positives

Vol 5. Planctomycetes, Spirochetes, Bacteroides, Fusobacteria

The examples of each group of bacteria are as follows:

*Vol 1. Archaea, cyanobacteria, phototrophs and deeply branched genera*

This volume has 3 important groups out of which one is in different domain (Domain - Archaea).

*1. Archaea*

Four important sections are present in the archaea.

a. Hyperthermophiles Ex. *Thermococcus, Sulfolobus, Thermosphaera*

b. Methanogens Ex. *Methanobacterium, Methanococcus, Methanosarcina*

c. Halobacteria Ex. *Halobacterium, Halococcus, Natronomonas*

d. Thermoplasma Ex. *Thermoplasma*

*2. Cyanobacteria*

They are the filamentous, oxygenic photosynthetic bacteria. They have special cells called *heterocyst* in which nitrogenase enzyme is present. The nitrogenase enzyme is responsible for fixing atmospheric $N_2$ into ammonia. Some cyanobacteria are single celled and some are non-heterocystous. Important groups are as follows:

1. Single celled cyanobacteria Ex. *Chrococcus, Gleotheca, Gleocapsa*
2. Filamentous non-heterocystous cyanobacteria Ex. *Oscillatoria, Lyngbya*
3. Filamentous heterocystous cyanobacteria Ex. *Anabaena, Nostoc, Tolypothrix*

*3. Anoxygenic phototrophs*

Single celled, sulphur required bacteria. They use $H_2S$ as electron donor. Ex. Green sulphur bacterium *Chlorobium*.

*Vol 2. Proteobacteria*

This volume has gram negative bacteria. They are further divided into 5 subgroups as α, β, γ, δ and ε. They contain medically, industrially and agriculturally important bacteria.

| S. No. | Imp. Bacteria | Characters | Example |
|---|---|---|---|
| α *Proteobacteria* | | | |
| 1. | Purple bacteria | Anoxygenic Photosynthetic - sulphur bacteria | *Rhodospirillum, Rhodobacter, Chromatium* |
| 2. | Associative Nitrogen fixing bacteria | These bacteria present in the rhizosphere of graminaceous plants and symbiotically fix atmospheric nitrogen. | *Azospirillum* |
| 3. | Symbiotic Nitrogen fixing bacteria | Form nodules in legume roots and fix atmospheric nitrogen. | *Rhizobium, Bradyrhizobium,* |
| | | Some form galls in the roots | *Agrobacterium* |
| 4. | Free living Nitrogen fixing bacteria | Present in the soil as heterotrophs - use verity of carbon sources in soil and fix atmospheric nitrogen | *Azotobacter, Beijerinkia* |
| 5. | Pseudomonas group | Some are Plant Growth Promoting Rhizobacteria | *Pseudomonas* |
| | | Some are pathogens | *Xanthomonas* |
| | | Some produce alcohol | *Zymomonas* |
| 6. | Rickettsia | Endoparasites | *Rickettsia* |
| 7. | Sulphur oxidizing bacteria | Uses S as electron donor - Chemolithotrophs - strict aerobes | *Thiobacillus* |
| 8. | Acetic acid producing bacteria | Fermentative bacteria | *Acetobacter, Gluconobacter* |
| 9. | Budding bacteria | Reproduction by budding like yeast | Caulobacter |
| 10. | Hydrogen bacteria | Hydrogen producing bacteria | *Alkaligenes* |

*Contd...*

| β Proteobacteria | | | |
|---|---|---|---|
| 1. | Nitrifying bacteria | Chemolithotroph - strict aerobe - soil bacteria - important form N cycle | *Ammonia to nitrite - Nitrosomonas Nitrite to nitrate - Nitrobacter* |
| 2. | Neisseria & relatives | | Neisseria |
| 3. | Spirillum | Aerobes & facultative aerobes | *Spirillum sp.* |
| 4. | Sheathed bacteria | | *Sphaerotilus* |
| γ Proteobacteria | | | |
| 1. | Purple sulphur bacteria | Anoxygenic photosynthetic - sulphur bacteria | *Thiobacillus, Thiospirillum* |
| 2. | Methylotrophs | Uses methane and methanol as carbon source | *Methylomonas, Methylobacter, methylococcus* |
| 3. | Coliforms | Present in the intestinal track of mammals | *Escherichia, Salmonella* |
| δ Proteobacteria | | | |
| 1. | Sulphur reducing bacteria | Anaerobes - use S as terminal electron acceptor | *Desulfovibrio, Desulfomonas* |
| 2. | Glding bacteria | Gliding movement | *Myxobacteria* |
| 3. | Vibrio group | Most are pathogenic | *Vibrio, Erwinia* |
| ε Proteobacteria | | | |
| | *Camphylobacter, Thiovulum, Thiomicrospora, Heliobacter* are common examples | | |

## *Vol 3. Low G+C gram positives*

| S. No. | Group | Characters | Example |
|---|---|---|---|
| 1. | Clostridia group | Strict anaerobes – mostly fermentative nutrition – few thermotolerant – endospore producers | *Clostridium, Thermoanaer obacterium, Thermoanaerobium* |
| 2. | Mycoplasma group | Absence of cell wall | *Mycoplasma, Mesoplasma, Spiroplasma* |
| 3. | Bacilli and Lactobacilli group | Lactic acid producing bacteria – endospore producers – aerobes – aerotolerant – fermentative nutrition | *Leuconostoc, Lactococcus, Streptococcus* |

## *Vol 4. High G+C gram positives*

| S. No. | Group | Characters | Example |
|---|---|---|---|
| 1. | Actinomycetes | Filamentous – sporangiospores – conidiospores – soil habitat – antibiotics producers | *Actinomyces, Nocardia, Sreptomyces* |
| | | Symbiotic with *Casuarina* – form root nodules – $N_2$ fixation | *Frankia* |
| 2. | Mycobacterium group | Presence of mycolic acid in the cell wall – acid fast staining – human pathogens | *Mycobacterium lepri* |
| 3. | Corynebacterium group | Human pathogens | *Corynebacterium diptheriaea* |

*Vol 5. Plancomycetes, Spirocheter, Bacteroides and Fusobacteria*

| S. No. | Group | Characters | Example |
|---|---|---|---|
| 1. | Chlamydia group | Obligate parasites to man, animal and birds | *Chlamydia* |
| 2. | Bacteroides | Obligate anaerobes | *Bacteroides* |
| 3. | Spirochete | Gram negative - flexile - endoflagella presence | *Spirocheta, Leptospira* |

CHAPTER

6

# Fungi

Eukaryotic, spore - producing, achlorophyllous organisms with absorptive nutrition that generally reproduce both sexually and asexually and whose thallus is usually filamentous, branched somatic structures known as hyphae, typically surrounded with cell wall.

Mycology - Study of fungi

The fungi can be divided into two groups namely

*True fungi*: Chytridiomycota, Zygomycota, Ascomycota, Basidiomycota

*Additional phyla*: Myxomycota, Dictyosteliomycota, Acrasiomycota, Plasmodiophoromycota, Oomycota, Labyrinthulomycota, Hypochitridomycoya.

## Economic Importance of Fungi

Following are the benefits of the fungi:

| S. No. | Benefits | Process | Example |
|---|---|---|---|
| 1. | Biotran-formations | Conversion of carbohydrate rich compounds to value added products like alcohol, steroids, bread etc. | *Yeast* (alcohol & bread) |

*Contd...*

| | | | |
|---|---|---|---|
| 2. | Useful metabolites | Antibiotics, plant growth regulators, hormones etc. | *Penicillium* (Penicillin) *Fusarium* (Gibberellic acid) |
| 3. | Enzymes | Industrial production of enzymes | *Aspergillus* (Amylase) |
| 4. | Biomass | Food and feed additives | *Agaricus* (Food) Yeast (Feed additive) |
| 5. | Biological control | Control the insects, other fungal pathogens, nematodes etc. | *Tichoderma* (Against soil plant pathogenic fungi) |
| 6. | Soil fertility | Nutrient cycle, decomposition of organic matter | Most of the soil fungi |

Following are the harmful effects of fungi:

| S. No. | Harmful effect | Cause/Symptom | Example |
|---|---|---|---|
| 1. | Food spoilage | Make the foods unable to use | *Aspergillus* (Stored grains) *Mucor* (Bread) |
| 2. | Mycotoxins (referred as the toxins produced by fungi on certain plant materials that will either consume directly as food or feed) | Causes diseases, tumors to the host animals | *Aspergillus flavus* (Groundnut) |
| 3. | Human & animal diseases | ill health, diarrhea, skin diseases | |
| 4. | Plant diseases | Root rot, rust, powdery mildew, blight, leaf spot, smut, etc. | *Phytophthora infestans* (Late blight of potato) |

## General characters of fungi

- Heterotrophic organisms devoid of chlorophyll
- They resemble plant by
  - cells wall
  - non-motile
  - reproduce by spores
- Separate from plants
  - No leaf, stem, root
  - No chlorophyll
  - Carbohydrate storage is glycogen instead of starch
  - Filamentous
- *Hypha* : Body of fungi ; microscopic tubular thread like hyphae branch in all directions. Collectively mass of hyphae is referred as mycelia (Singular : mycelium). Some don't have these structures. Ex. Yeast
- They use variety of carbon sources like cellulose, starch, hemicellulose etc.
- Temperature optima : 25-30° C ; Max : 40° C Min : 10° C some thermophilic fungi prefer to grow at 50° C
- pH optima : 4-7
- *Oxygen* : Majority of them aerobes; Some facultative anaerobes (Ex. Yeast), and some obligate fermentative type (Ex. chytridomycota)
- *Light* : Not required

*Dimorphic* : Some organisms can exist either hyphae or single celled. Ex. Disease cause fungi of human and animals. They exist as hyphae when outside the host and in the host, as single celled. They are referred as dimorphic fungi.

## Morphology of fungi

*Hyphae:* thin, transparent, tubular wall - filled with protoplasm.

The cross walls are referred as septa (Septum).

If septa is present, the hyphae is referred as *septate hyphae* and absence of septa, *aseptate / non septate / coenocytic hyphae*.

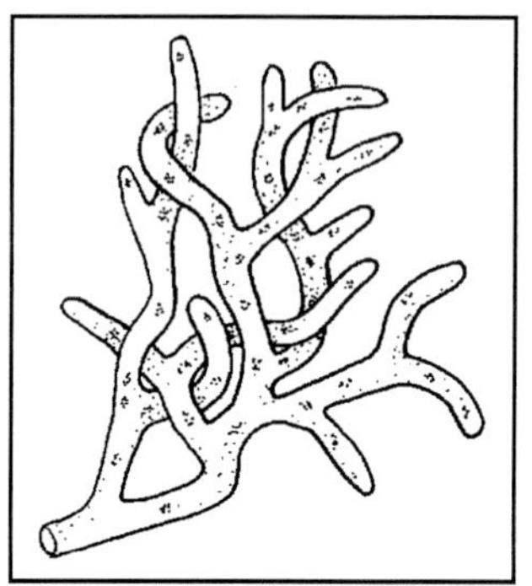
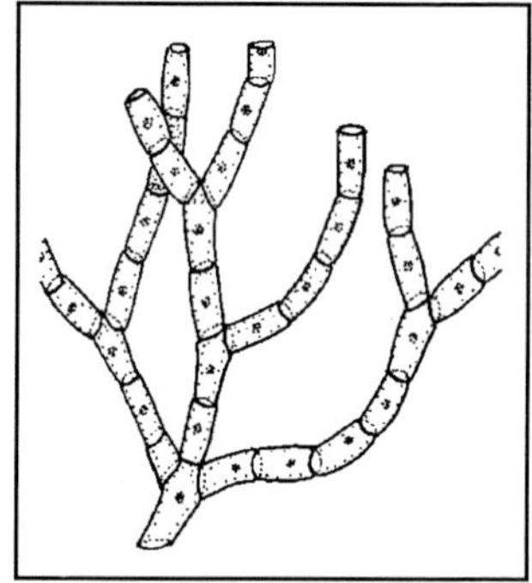

Aseptate and septate mycelium

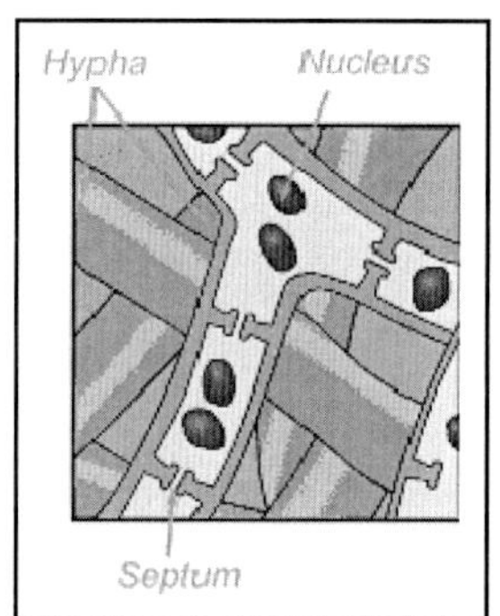

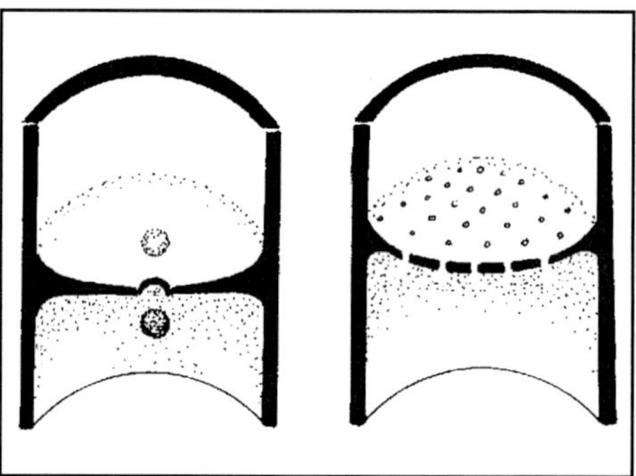

Cross section of hyphae and septum

*Cell wall*

The cell wall of fungi is made up of cellulose, glucans, chitins etc.

Glucan: glucose with $\alpha$ - 1,3 ; $\alpha$ - 1,6; $\beta$ - 1,6 and $\beta$ - 1,4 linkage

Chitin: b - 1,4 linked N- acetyl glucosamine

Oomycetes – cellulose is the major compounds of cell wall

Normally the hypha / mycelium are unnoticed in underground nature which cannot see by eyes. The mycelium is always embedded in the soil/ host plant or animal. But sometimes they become organized to form large structures to see by naked eyes, which are referred as special vegetative structures. Following are some of modification of vegetative structures of saprophytic fungi.

1. *Stroma:* compact, somatic structure much like a miniature mattress or a cushion formed are referred as stroma.

2. *Sclerotia:* Hard resting bodies resistant to unfavorable conditions Ex. *Sclerotium* sp. and *Claviceps purpurea*

3. *Rhizomorphs:* Thick strands of hyphae, consolidated tissue like are referred as rhizomorphs. They become dormant at adverse condition and useful for nutrient absorption.

Following are the modification of vegetative structures of parasitic fungi.

1. *Haustoria:* The fungi penetrates the host cells by branched hypha and produce bulbous structures are called as haustoria. This structure is used to absorb the nutrients from host plant cell. Ex. Rust fungi

2. *Appressoria:* Infection structures formed at the tip of germ tubes or hyphae outside the host. The appressoria is the spore bearing structure of parasitic fungi. Ex. *Puccinia, Peranospora*

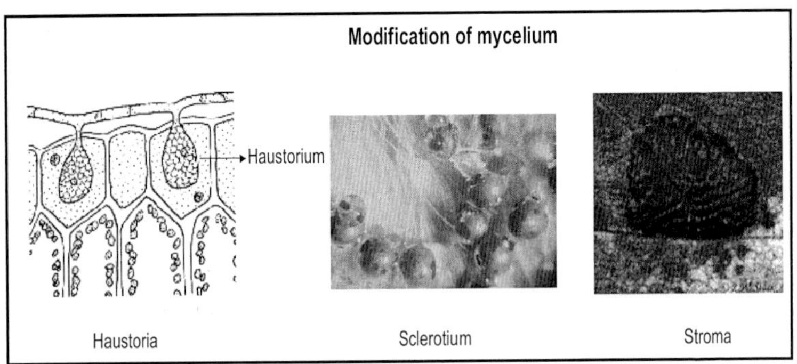


## Nutrition

Fungi perform the absorptive type of nutrition. They derive food by four different ways as follow:

1. *Saprophyte :* Uses dead organic matter to derive food – Heterotrophs Ex. *Mucar, Aspergillus, Penicillium*
2. *Parasites :* Uses living tissues to derive food. The parasitic fungi were further divided into three groups as follows based on their mode of nutrient absorption from host.
   - Perthotroph/Necrotroph: Use enzyme or toxin to kill the host cell and derive the food from dead cells
   - Biotrophs: Produce special vegetative structures (haustoria) and absorb the nutrients from live cells. They are ecologically obligate parasites
   - Hemibiotrophs : Initially require living host and later uses the dead cells Ex. *Colletotrichum lindemuthianum*
3. *Symbiosis :* Derive food from the host by beneficial each other. Ex. Lichens - Algae and fungi sybmiosis; Mycorrhiza- Higher plant roots and fungi symbiosis
4. *Predation:* Animal trapping fungi capture the small animals like worms, protozoa etc. and kill the cells and derive their food. Ex. *Arthobotrys, Dactylella*

*Ecological groups*

The fungi can be divided into many groups based on the ecological conditions.

| S. No. | Group | Habitat | Example |
|---|---|---|---|
| 1. | Soil fungi | Soil | *Penicillium, Aspergillus* |
| 2. | Lignicolous fungi | Woody tree barks where rich of lignin | *Polystictus* |
| 3. | Entomogenous fungi | Insects | *Endomopthora* |
| 4. | Coprophilous fungi | Cowdung | *Pilobolus* |
| 5. | Aquatic fungi | Water | |
| 6. | Cellulolytic fungi | Cellulose rich soil | *Chaetomium* |
| 7 | Dermatophytes | Skin | *Trichophyton* |

## Reproduction of fungi

Fungi can reproduce by vegetative, asexual or sexual reproduction.

*Vegetative reproduction*

1. *Fragmentation :* The hypha fragment to many cells and each cell will raise as new individual. Ex. Yeast.
2. *Fission :* As bacteria, binary fission gives two new daughter cells. Ex. Yeast.
3. *Budding :* In yeast, the cell forms a bud like appendage, in which the divided nucleus will move and new cell formed. The detached bud will raise to new individual.
4. *Sclerotia :* resting vegetative structure can give the new individual.
5. *Rhizomorph :* also gives the new individual.

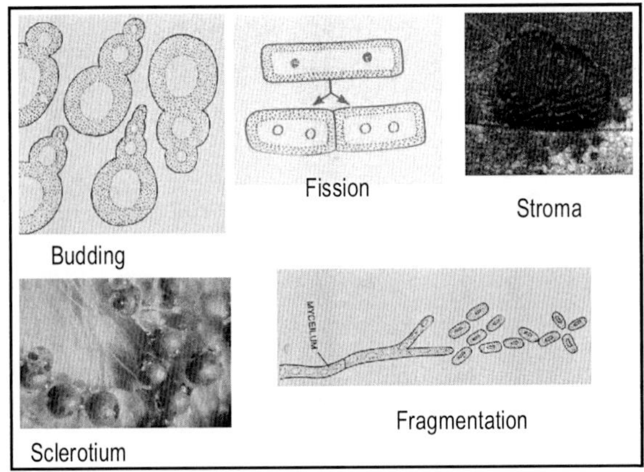


Vegetative reproduction of fungi

*Asexual reproduction*

1. *Arthospores/Oidiospores :* The broken hypha behave as spores are referred as arthospores or oidiospores.

2. *Chlamydospores :* The cells become resting body and separate from hyphae referred as chlamydospores. When favourable condition araises, the chlamydospore will form new mycelium Ex. *Fusarium.*

3. *Asexual spores :* Sporangiospores -spores produce from sporangium (sac like organ); Conidia - Spore produced at the tip of the hyphae.

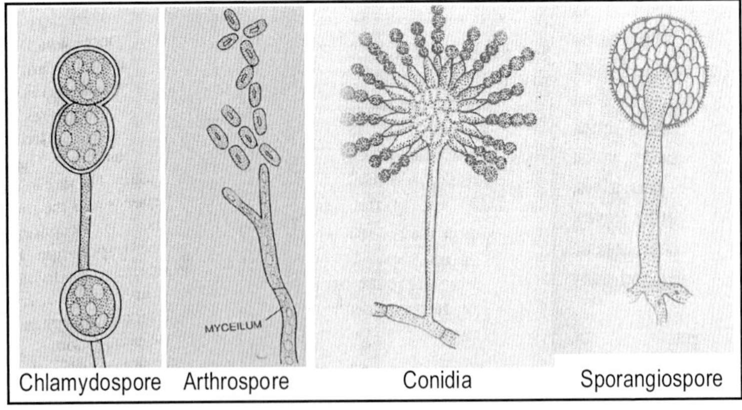

*Exogenous & Endogenous spore*

Spore produced inside the thallus is referred as endogenous spore (Ex. Sporangiospore) and spore produced outside the thallus is referred as exogenous spore (Ex. Conidia).

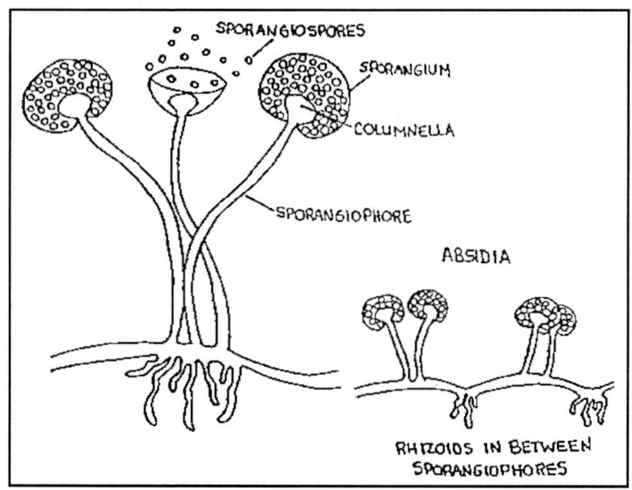


Endogenous spore - Sporangiospore of Rhizopus

*Motile & Non-motile spores*

Spores with flagella are referred as *Planospores*. The spores will have whiplash and tinsel type flagella either or both. Ex. Zoospores.

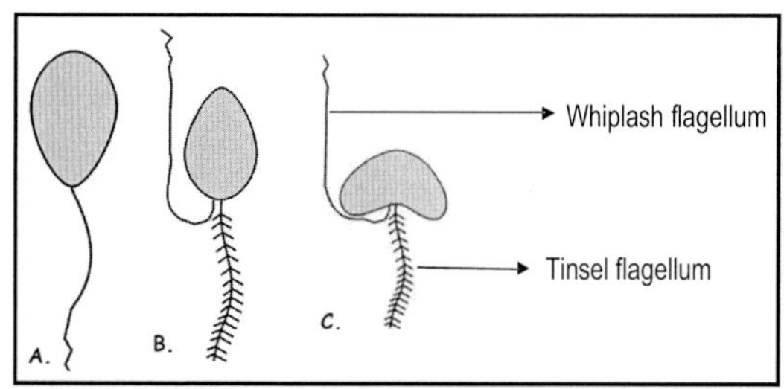


[illegible]
[illegible]
[illegible]

Spores without flagella are referred as *Aplanospores*. Ex. Sporangiospores, conidia.

Apart from these basic type of spores, pynidia, pycnia, acervuli, sporodochia are some other asexual spores produced by Basidiomycetes fungi.

## Sexual reproduction

Sexual reproduction of fungi involves the following sequence of processes.

1. *Plasmogamy* : Union of protoplasm of two thallus or gametes.
2. *Karyogamy* : Union of two nucleus
3. *Meiosis* : Cell division to maintain the n level of chromosome.

Generally, all the thallus are at 'n' level of chromosome, referred as *haplophase* and before meiosis, the chromosome level is '2n', which is referred as *diplophase*.

After meiosis spores were formed, which vary with different classes of fungi. The sexual spores are Oospores; Zygospores; Ascospores and Basidiospores

The sex organs of fungi are referred as *Gametangia / gametangium*

The Cell / nucleus involved in the sexual reproduction is *gamete* and the fusion of two gametes lead to sexual reproduction ie., plasmogamy, karyogamy and meiosis.

*Isogametangium & Isogametes*

Male & female sex organs and sex cells cannot be differentiate morphologically.

*Anisogametangium & Anisogametes*

Male and female sex organs and sex cells (gametes) structurally differ each other is referred as Anisogametangium and anisogametes.

Male gametangium - Anthridium

Male gamete - Anthrozoid

Female gametangium – Oogonium, Ascogonium

Female gamete - Oosphere

Union of two isogametes is referred as *isogamy* and union of anisogametes is referred as *anisogamy*.

1. *Monoecious / Hermaphrodite :* Male and female sex organs present in same thallus
2. *Dioecious :* Male and female are present on different thalli.

*Homothallic fungi*: Male and female gametangia are present in same thallus and can reproduce by itself referred as homothallic fungi. Also referred as "Self – fertile".

*Heterothallic fungi:* Male and female gametangia are present in different thallus and cannot reproduce by itself referred as heterothallic fungi. Also referred as "Self – sterile" - need another type of thallus for sexual reproduction

*Pseudohomothallic fungi*: They are heterothallic fungi (self sterile) but their spores are self fertile referred as pseudohomothallic fungi.

## Holocarpic & Eucarpic Fungi

If whole thallus/hyphae used or converted as the reproductive phase, the fungi is referred as *holocarpic* fungi. Ex. Yeast. Reproductive organ arise from the thallus referred as *eucarpic* fungi. Ex. *Aspergillus*. The sexual stage of these fungi are referred as teleomorph; asexual stage is anamorph and collectively referred as holomorph.

## Common Sexual Reproductions

*1. Planogametic copulation*

Fusion of naked motile gametes is referred as planogametic copulation. Ex: *Plasmodiophora* "Planogametes" - motile gametes.

*2. Gametangial contact*

Two gametangia come in contact but not fuse. The male nucleus (gamete) migrates through a pore or fertilization tube to female gametangium and sexual reproduction takes place. Ex. *Aspergillus, Peniicillium.*

*3. Gametangial copulation*

Two gametangia (male and female) fuse and sexual reproduction takes place, referred as gametangial copulation. Ex. Zygospore of Zygomycetes - *Mucar, Rhizophus.*

*4. Spermatization*

A uninucleate nonmotile spore-like male structures acted as "male gamete" (Spermatium) (gametangium: spermagonium / antheridium) which pass through female gemetangium (Oogonium) leads fertilization. This process of sexual reproduction is referred as spermatization. Ex: *Neurospora.*

*5. Somatogamy*

The sex organs are not produced. The somatic cells (thallus) take part in the sexual fusion is referred as somatogamy. Ex. *Morchella*

*Parasexuality:* The sexual reproduction (plasmogamy - karyogamy - meiosis) occur in sequence but not at a specific points in the life cycle of an individual is referred as parasexuality. Ex. *Aspergillus nidulans.*

## Fungal classification

The recent advances in fungal classification includes:

- Recognition of artificial nature of three / five kingdom classification
- Theory of data analysis techniques of phylogenetic systems.
- Molecular techniques in mycology
- Discovery of new fossils and taxa

Present classification is *"Phylogenetic classification"* based on evolutionary relationship. The fungi can be grouped in to two groups based on their ancestors.

Monophyletic groups: Group of individuals that contain a common ancestor and all its descendants.

Polyphyletic groups : Group of individuals that donot share a common ancestor.

The fungi are divided into 3 major groups

*1. Kingdom : Fungi (one monophyletic group)*

*Note:* The common name fungi and kingdom *Fungi* are differentiated by writing as scientific name – italic or underline

- Phylum : Chytridiomycota
- Phylum : Zygomycota
- Phylum : Ascomycota
- Phylum : Basidiomycota

*2. Kingdom – Stramenopila (one monophyletic group)*

- Oomycetes
- Hyphochytriomycota
- Labyrinthulomycota

*3. Protists (All are polyphyletic groups)*

a) Plasmodiophoromycota

b) Dictyosteliomycota

c) Acarysomycota

d) Myxomycota

The protists are formerly referred as "Slime molds".

*Kingdom : Fungi*

*Phylum : Chytridiomycota*

- This phylum has single class- *chytriomycetes.*
- Only members of fungi produce motile cells at some stage of their life history.
- Asexual spores - Zoospores.
- Sexual reproduction - Gametangial copulation.
- Coenocytic structure of hyphae with well developed mycelium.
- Eucarpic.
- Rhizoids / rhizomycelium - well developed used for nutrition absorption and for anchor.
- These fungi are referred as "Chitrids".
- Occur in aquatic, soil, cattle rumans etc.
- Many orders , families are present.

Type genera : *Chytriomyces, Allomyces, Synchytrium.*

*Phylum : Zygomycota*

This phylum includes two classes namely Zygomycetes and Trichomycetes.

*Class : Zygomycetes*

- Characterized by thick walled resting spores called "Zygospores" – developed through zygosporangium.
- All individuals produce this sexual spore.
- Morphological variation is there.
- Coenocytic mycelium.
- Asexual reproduction- sporangiospores (Aplanospores).
- Sexual reproduction : Zygospores

  • Gametangial copulation leads to formation of Zygospore, which is a resting spore. When favorable condition, zygospore germinate gives the sporangiospores.

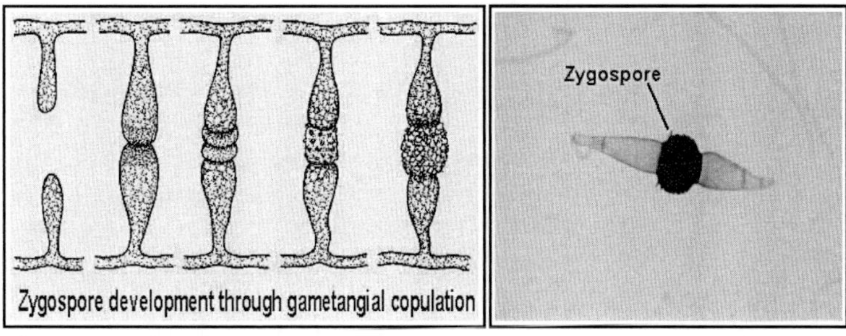

Zygospore development through gametangial copulation

- Disease causing organisms.
- Food borne pathogens.
- Food spoilage.
- Type genera : *Mucar* , *Rhizopus* , *Entomophthora.*
- Important orders are Mucarales; Entomophthorales; Zoopagales.

*Class : Trichomycetes*

- Obligate associated with insects, millipedes, arthropods.

- symbiotic association occurs.
- Most of them grown internally within the gut of these animals referred as 'holdfast'.
- Asexual reproduction.
- Amoeboid cells.
- Arthrospores.
- Sporangiospores.
- *Trichospores :* These are exospores with single uninucleated spores exclusively present in this class fungi.
- *Sexual reproduction:* Zygospores.

*Type genera : Harpella, Smittium, Amoebidium, Asellaria, Enterobryus.*

*Phylum : Ascomycota*

The characters of Ascomycota fungi are

- Septate mycelium.
- Sometimes dikaryotic mycelium (2 nuclei per cell).
- Ascospores – sexual spores - 4 or 8 ascospores in a sac – called "Ascus".
- Isogametangial contact, anisogametangial contact, spermatization and somatogamy are the common sexual reproduction methods.
- Fruiting body of Ascospores is "Ascocarp".
- Few don't produce ascocarp Ex. Yeast.
- *Asexual spores :* Conidia / sporangiospore-Aplanospores.
- Variety of shapes of conidia are present.

- Few individuals have only asexual stage and few have both sexual and asexual stage.

The fungi which have both sexual and asexual stage are referred as *perfect fungi* and the fungi having the asexual stage alone are referred as *imperfect fungi.*

The structure of ascocarp varies with three different types.

1. *Cleistothecium* : Closed ascocarps are called as cleistothecium.
2. *Perithecium* : Closed ascocarp but a pore or opening is present to facilitate the release of ascospores are called as perithecium.
3. *Apothecium* : Opened ascocarp is referred as apothecium.

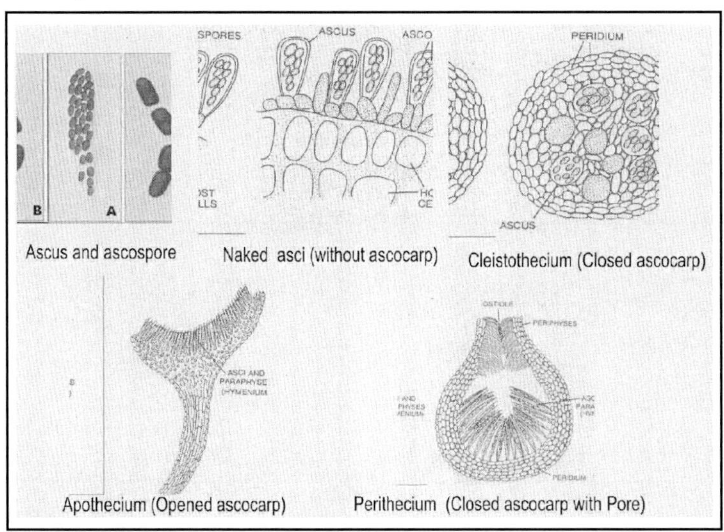

Ascus and ascospore    Naked asci (without ascocarp)    Cleistothecium (Closed ascocarp)

Apothecium (Opened ascocarp)    Perithecium (Closed ascocarp with Pore)

## Different Classes of Ascomycota

1. *Asexual ascomycetes* : Deuteromycetes/Imperfect fungi only asexual reproduction occurs by means of sporangia / conidia. Ex. *Penicillium, Paecilomyces, Metarrhizium.*

2. *Archiascomycetes:* Saprophytic & parasitic fungi. Ex. *Taphrina, Schizosaccharomyces*
3. *Order:* Saccharomycetales. Yeast: 4 Ascospores per ascus; Ex. *Saccharomyces, Candida, Kluyveromyces*
4. *Order :* Eurotiales (filamentous ascomycetes). Ascospores with cleistothecium type ascocarp Ex. *Aspergillus, Paecilomyces, Talaromyces*
5. *Ascomycetes with perithecia (Pyrenomycetes) :* Perithecium type ascocarp. Ex. *Neurosposa , Fusarium, Claviceps*
6. Ascomycetes with Apothecium (Discomycetes). Ex. Lichens, Mycorrhiza

*Lichens :* Association of fungus with algae / cyanobacteria in which a single thallus formed are referred as lichens. The fungal component is referred as mycobiont and algal component is photobiont. Ex: *collema, Parmelia.*

Ecological groups of lichens:

| | | |
|---|---|---|
| Rock surface | : | Saxicolous lichens |
| Inside rock | : | Endolithic lichens |
| Soil | : | Terricolous lichens |
| Bark | : | Corticolous lichens |
| Leaves | : | Folicolous lichen |

*Mycorrhiza (Myco – fungi; rhiza – roots)*

The association between fungi and higher plant roots is referred as mycorhhiza. They are divided into two groups as endomycorrhiza, which penetrate the root cell and symbiotically beneficial with roots; ectomycorrhiza, which colonize on the surface of the higher plant roots and symbiotically beneficial with roots. Ex. *Glomus, Gigaspora* - Endomycorrhiza; *Amanita, Bolytus* - Ectomycorrhiza.

7. Ascomycetes with Ascostromata: Production of asci in the stroma (cushion like structure) referred as ascostroma is the typical character of this class fungi. Ex:*Venturia, Podonectria.*

8. Other filamentous ascomycetes-Powdery mildew disease causing fungi. Ex. *Erysiphae, Leveillula.*

*Phylum: Basidiomycota*

The characters of basidiomycetes fungi are as follows:

- *Sexual spore :* Basidiospores
- *Spore forming structures :* Basidium
- *Fruiting body* : Basidiocarp
- 4 basidiospores per basidium formed
- There are three different stages of life cycle such as a. primary mycelium - septate - single nucleated b. Secondary mycelium - septate- heterokaryons stage and c. Tertiary mycelium in which basidiocarp formation
- *Asexual reproduction:* Oidium / oidiospores

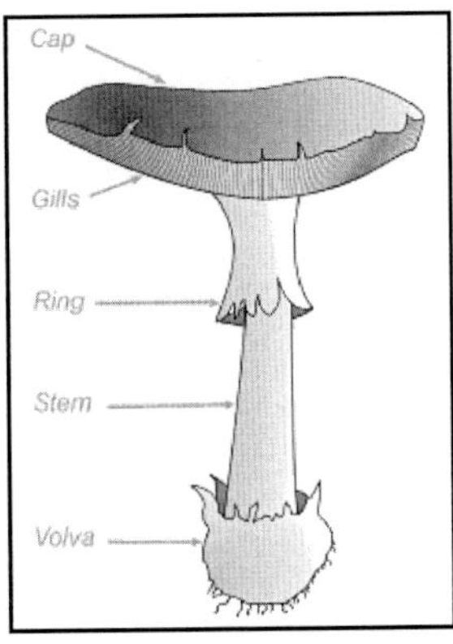

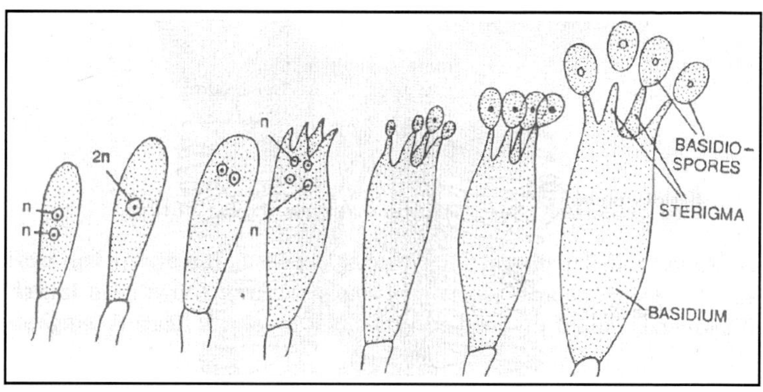


Basidiospore & Basidium development

Following are the sub groups of basidiomycetes

1. *Order :* Agaricales (Mushrooms) Ex. *Agaricus, Coprinus.*
2. *Class:* Gasteromycetes : Puffballs , earth stars , Birds nest fungi.
3. *Order:* Aphyllophorales : *Polyporus* , Chantarelles , tooth fungi , coral fungi, corticoides.
4. *Order :* Uridinales : rust fungi, *Puccinia, Elromyces.*
5. *Order :* Ustilaginales : Smut fungi *Ustilago, Tilletia, Urocystis.*
6. Other basidiomycetes Ex: *Tremella.*

*Kingdom : Stramenopila*
*Phylum : Oomycota*

- *Class:* Oomycetes.
- *Sexual:* Oospores.
- *Asexual:* Zoospores.
- *Type genera : Pythium, Phytophthora, Perenospora, Plasmopara, Albugo.*

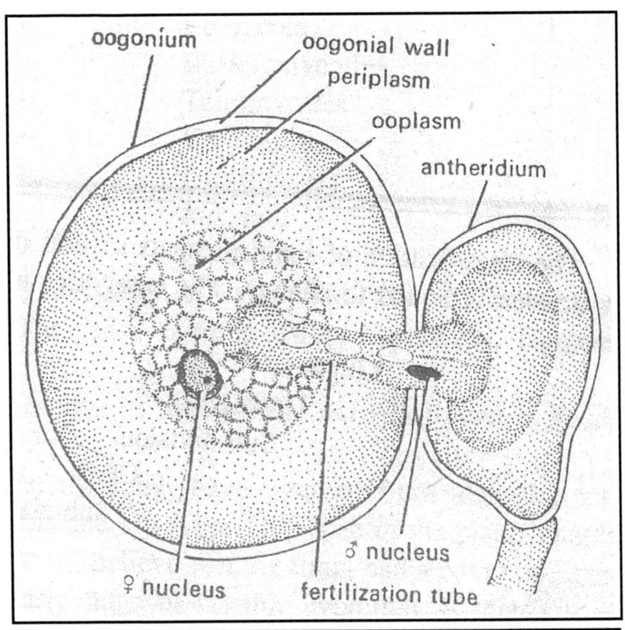


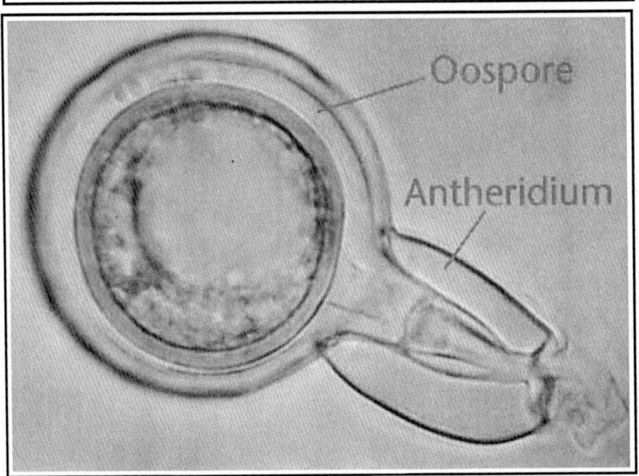


Oospore development through gametangial contact

*Phylum : Hypochitridomycota*

- Holocarpic fungus
- Type genera : *Rhizidiomyces apophysatus*

*Phylum : Labyrinthulomycota*

- Net slime molds.

*Protists*

These group of organisms are not true fungi. They don't have mycelial structure as that of true fungi. They are phylogentically primitive to fungi and have the characters of fungi and protozoa.

*Phylum: Plasmodiophoromycota*

- Endoparasitic slime molds
- The thallus is referred as plasmodia. They are acellular, mass of protoplasm in which the nuclei are present. Ex. *Plasmodiophora.*

*Phylum: Dictyosteliomycota*

- Cellular slime molds Ex. *Dictyostelium.*

*Phylum: Myxomycota*

- True slime molds : *Myxcomycetes skemonitis.*

CHAPTER

7

# Algae

Algae are a large and diverse group of eukaryotic organisms that contain chlorophyll and carry out the oxygenic photosynthesis.

Study of algae is referred as phycology or algalogy

They vary from cyanobacteria or blue green algae by their eukaryotic nature.

## Occurrence

The algae are mainly present in aquatic ecosystems.

They are present in

- Oceans, sea, salt lakes, fresh water lakes, ponds and streams.
- Damp soils, rocks, stones, tree barks etc.
- Small aquatic free floating microscopic algae in water are referred as planktons or phytoplanktons.
- Some are present as parasites to plants.

## Economic Importance

Following are the beneficial effects of algae:

- The algae are the primary producer of food chain of aquatic ecosystem.

- They are also the $CO_2$ fixers.
- Algal cell wall contains several economic compounds such as alginic acid, agar, carrageenan.

*Agar* : is a polysaccharide obtained from the red alga – *Gelidium*. It is used as solidifying agent for media preparation in microbiology labs; cheese, jelly preparations and as packing for canned foods.

*Alginic acid* : is obtained from brown algae *Macrocystis, Agarum, Laminaria*. It is useful in smooth consistency of ice cream, tooth impressions of dental labs, jelly preparations, immobilization matrices using calcium alginates.

*Carrageenan* : Obtained from *Chondrus, Gigartina*. It acts as stabilizer of foods and ice creams, binder of tooth paste, finishing compound of textile and paper industry, thickening agent of shaving creams, soaps etc.

- Diatoms are the group of algae which contain silica in cell wall. The diatomaceous earth is used for filter preparations.
- *Porphyra* (a red algae) used as food (Japanese uses this alga for Nori preparations)
- *Chlorella* (green algae) used as feed additive for domestic animals
- *Spirulina* (green algae) has many medicinal properties
- The different group of these algae is used for pigments production such as carotenoids, xanthophylls, phycocyanins etc.

Following are some harmful effects of algae:

- *Prototheca* alga produces bursitis to mammals and human (Inflammation of joints)
- *Cephaleurous parasiticus* causes red rust disease to tree crops such as tea, coffee, mango and sapota.

- Aquatic algae produce neurotoxins which are of highly poisonous. (Ex. *Gymnodium*).

## Morphology

Wide range of size and shapes are occurred in algae.

1. *Unicellular :* single cells, motile or nonmotile Ex. *Chlamydomonas, Chlorella*
2. *Colonies :* Assemblage of individual cells with variable or constant number of cells that remain constant throughout the colony life Ex. *Diatoms, Volvox*
3. *Filaments :* daughter cells remain attached after cell division and form a cell chain; adjacent cells share cell wall (distinguish them from linear colonies) Ex. *Spirogyra, Spirulina*
4. *Coenocytic forms :* one large, multinucleate cell without crosswalls Ex. *Vaucheria*
5. *Parenchymatous and pseudoparenchymatous algae:* mostly macroscopic algae with tissue of undifferentiated cells and growth originating from a meristem with cell division in three dimensions; pseudoparenchymatous superficially resemble parenchyma but are composed of appressed filaments Ex. *Porphyra*
    - Eukaryotic nature
    - Presence of thin and rigid cell wall
    - Presence of chloroplasts

## Cell wall

- The algal cell wall is primarily made up of cellulose fibres. But, the cellulose fibres are modified by additional polysaccharides like xylans, mannoses, alginic acids etc.

- Some algal cell walls deposited with calcium carbonate, referred as calcarious or coralline algae.
- Some algae, the cell wall is absent. Ex. *Euglena*
- Silica deposited cell wall Ex. Diatoms. They are extreme resistance to decomposition.

## Nutrition of Algae

- Photosynthesis is the primary mode of nutrition (C uptake as $CO_2$).
- Photosynthetic algae are mostly considered as *photoautotrophic.*
- Numerous phototrophic algae can, however, take up dissolved organic matter or engulf bacteria and other algal cells as particulate prey; they are referred to as *mixotrophic.*
- Some algae cannot synthesize essential vitamins (biotin, thiamine, cobalamin = $B_{12}$) and have to import them: referred as *auxotrophic* species.

## Motility

Presence of flagella for motility for few motile algae Ex. *Euglena, Chlamydomonas*

Dianoflagellates have two flagella with different lengths.

## Pigments

Since algae are the photosynthetic organisms, their light harvesting system, by pigments varies among different group of algae. Basically the pigments are grouped into following categories:

1. Chlorophyll – a, b, c, d, and e - *lipid soluble*
2. Carotenoids – Carotenes and xanthophylls - *lipid soluble*

3. Biloproteins (Phycobilins) - Phycocyanin and phycoerythrin - *water soluble*

The concentration of these pigments varies with different algae leads to appearance of particular colour. Ex. Chlorophyll rich leads to green colour; phycoerythrin rich leads to red colour; and so on.

## Reproduction

1. *Asexual Reproduction* : One individual can produce copies of itself.

- *Binary fission* : many unicellular algae, longitudinal or transverse cell division.
- *Fragmentation*: colonies or filaments break into two to several pieces that continue to grow.
- *Akinetes* : Enlarged vegetative cell with thick wall and storage products; survival of harsh conditions rather than additional copies.
- *Zoospores / Aplanospores* : Zoospores are flagellated reproductive cells from which a new individual/ colony can grow; sometimes the spores begin to develop within the mother cell and lack flagella: aplanospores; aplanospores can develop into zoospores.

2. *Sexual Reproduction* : Plasmogamy, the fusion of haploid cells (gametes), preceeds karyogamy (nuclear fusion) to form a diploid zygote are the principle events of sexual reproduction.

- Gametes look like vegetative cells or very different.
- *Isogamy:* both gametes look identical.
- *Anisogamy:* male and female gametes differ morphologically.

- *Oogamy:* One gamete is motile (male), one is nonmotile (female).
- *Monecious:* both gametes produced by the same individual.
- *Diecious:* male and female gametes are produced by different individuals.
- *Homothallic:* gametes from one individual can fuse (self-fertile).
- *Heterothallic:* gametes from one individual cannot fuse (self-sterile).
- Gametophyte: typically haploid, produces gametes by mitosis.
- *Sporophyte:* typically diploid, produces spores by meiosis.

All types of reproductions studied in fungi will be applicable to algae too. *Note:* refer fungi for further details

## Taxonomic Groups

*Phylogeny of Algae*

- Algae did not evolve from one single, common ancestor but evolved at different times from different ancestors. (Polyphyletic group) so, Algae are, therefore, not monophyletic
- Some algae are closer related to heterotrophic flagellates or ciliates than to other algae
- *Eukaryotic algae:* possess a double membrane enclosed nucleus and usual other membrane-enclosed organelles (mitochondria, chloroplasts)
- *Origin of organelles:* symbiotic enclosure of free living bacteria or cyanobacteria, horizontal transfer of genes from endosymbionts to host nucleus.

The pigments are the main taxonomic character for grouping the algae in to different phyla. The following table shows different groups of algae with suitable examples.

| **Taxonomic group** | **Pigments** | | | **Examples** |
|---|---|---|---|---|
| | **Chloro-phyll** | **Carotenoids** | **Biloproteins** | |
| Rhodophycophyta (Red algae) | Chl a,d | β-carotene Zeaxanthene | Phycoerythrin Phycocyanin | *Polysiphonica Gelidium* |
| Xanthophycophyta (Yellow green algae) | Chl a,c,e | β-carotene heteroxanthene | - | *Vaucheria* |
| Chrysophycophyta (Golden algae) | Chl a,c | β-carotene Fucoxanthene | - | *Ochromonas* |
| Phaeophycophyta (Brown algae) | Chl a,c | β-carotene α-carotene Fucoxanthene Violaxanthin | - | *Kleps, Sorgassum natans* |
| Bacillariophycophyta (Diatoms) | Chl a,c | β-carotene | - | Diatoms |
| Chlorophycophyta (Green algae) | Chl a,b | β, γ-carotene | - | *Chlorella, Chlamydomo nas, Spirogyra* |
| Cryptophycophyta (Cryptomonads) | Chl a,c | α carotenes | Phycoerythrin Phycocyanin | *Cryptomonas* |
| Pyrrophycophyta (Dianoflagellates) | Chl a,c | β-carotene, Peridinin | - | *Noctiluca* |
| Euglenophycophyta (Euglenoids) | Chl a,b | - | - | *Euglena* |

## *Lichens*

Association of fungi with algae or cyanobacteria in which a single thallus formed are referred as lichens.

Fungal component is referred as *mycobiont* and algal component is *photobiont*.

## *Thallus*

Top layer of tightly woven mycelium, below which the photosynthetic cells and followed by another layer of fungus formed a complex structure which can not be distinguished the algae and fungi.

*Fungi* : Ascomycetes and few basidiomycetes; Algae – green algae.

About 18000 species of lichens have been identified. Based on their structure, they were divided into three groups as

1. *Crustose* : flat apprised
2. *Foliose* : leaf like
3. *Fructicose* : shrub like

*Reproduction:* by means of sordia (knot of fungal hyphae with few algal spores are referred as sordia).

*Economic importance of lichens*

1. Production of unusual compounds such as phenolics etc.
2. Commercial production of litmus pigment indicator
3. Commercial production of essential oils for perfumes
4. Indicator system form pollutants.

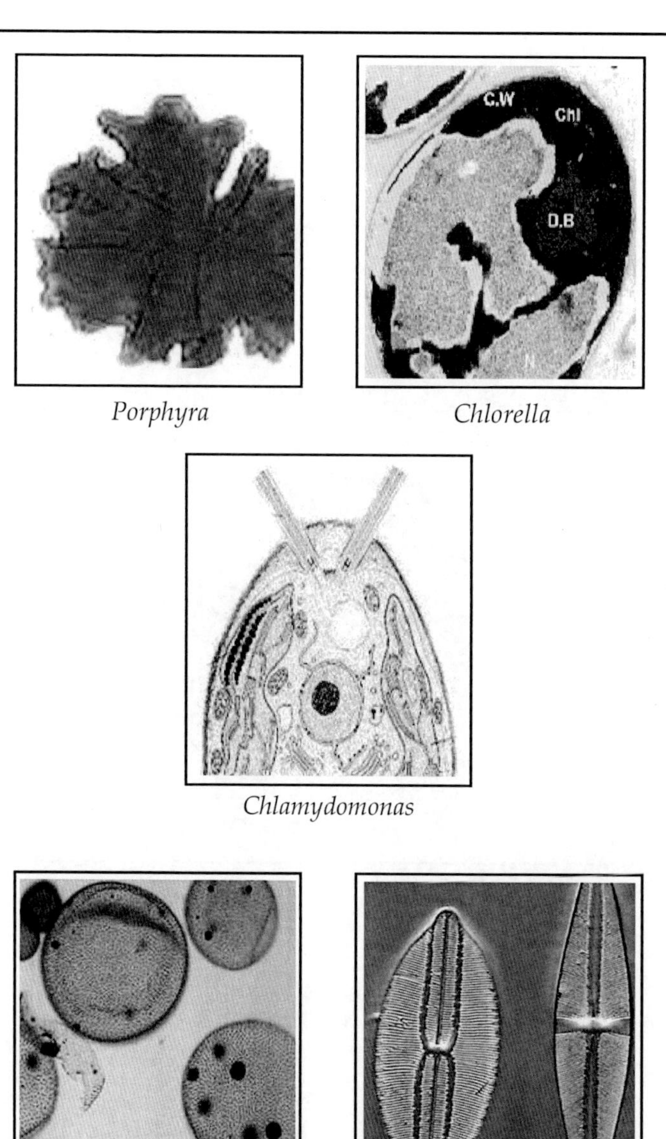


*Porphyra* *Chlorella*

*Chlamydomonas*

*Volvox* *Diatoms*

Some algae

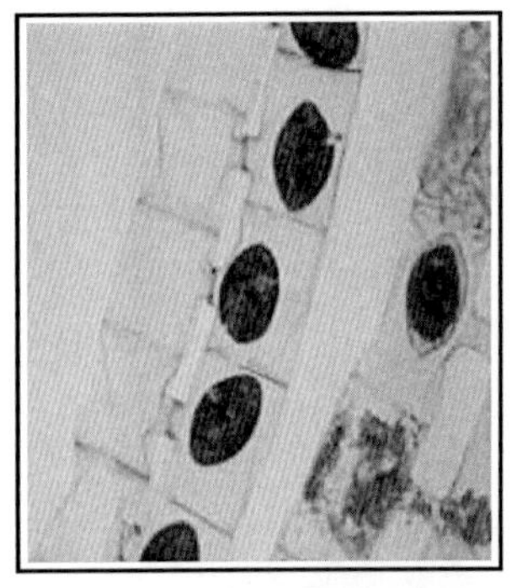

*Spirogyra*

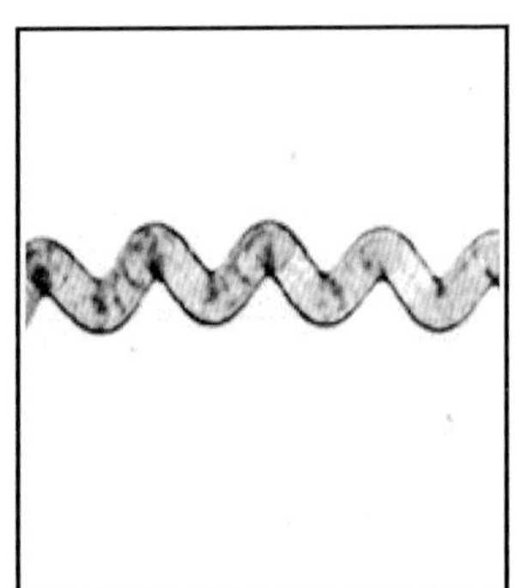

*Spirulina*

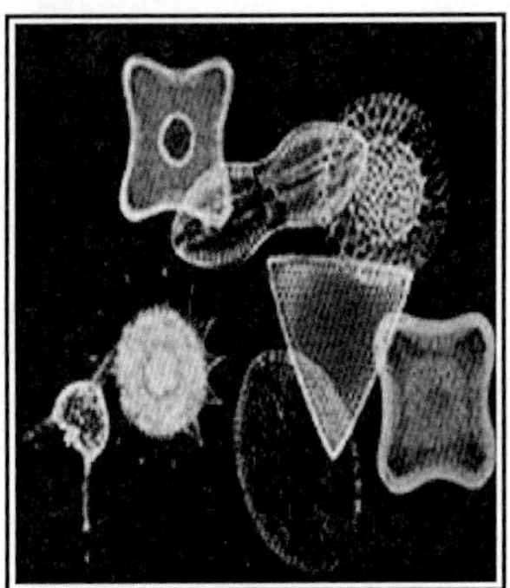

*Diatoms*

*Sargassum*

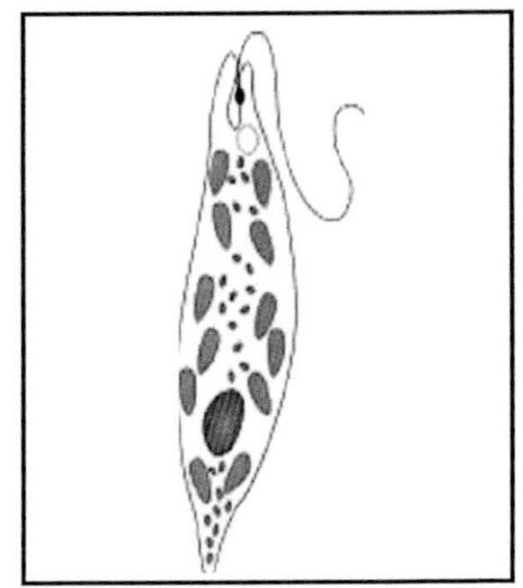

*Euglena*

CHAPTER 8

# Protozoa

Proto – first; Zoon – animal

Protozoa are single celled eukaryotic organisms, which lack cell wall and ability to move at some stage of their life cycle.

They are referred as the first animals.

There are about 65000 described species grouped into 7 phyla. Among them, 50 per cent are not available as living forms (as fossils) and out of remaining 50%, 22,000 species are free-living and 10,000 species are parasitic forms.

The protozoa are distinguished from other organisms by following characters

- From bacteria – Eukaryotic nature and greater size
- From algae – Lack of chlorophyll and cell wall
- From fungi – Motility and lack of cell wall
- From slime molds (Protists) - Lack of fruiting body

## Occurrence of Protozoa

- Present in all moist habitat
- Sea, soil and fresh water are common habitat
- Also present as parasites in higher animals

## Ecological Groups

Protozoa may be divided into three groups based on their ecological conditions namely,

1. *Free-living* : Occur in salt water, fresh water, sand, soil and decaying organic matter
2. *Symbiotic forms* : Occur in guts of termite and help in the digestion of woody materials. Ex. *Trychonympha* protozoa present in the guts of termites help to digest the cellulose of woods.
3. *Parasitic forms* : Occur in the body of higher animals and human and cause diseases.

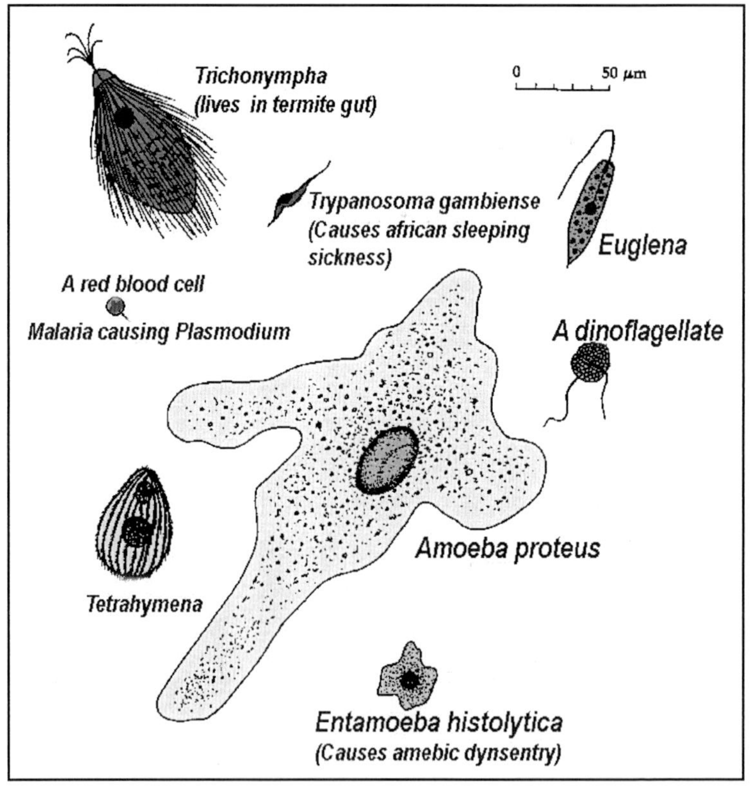


A Few Well-known Protozoa

## Morphology

Protozoa exhibit a wide variety of morphologies. There is no definite shape or morphology which would include a majority of protozoa. Shapes range from amorphous and ever-changing forms of amoeba to relatively rigid forms. Size and shape of these organisms show diversity.

Shape varies from amoeboid to all possible shapes.

Size ranges from 200 to 600 mm diameter protozoa are available.

*Plasmodium* – 2-5 mm; *Paramecium* – 200 - 500mm; *Spirostomum* – 3 mm;

Like eukaryotic cell, the protozoan cells consist of cytoplasm, nucleus, endoplasmic reticulum and all regular cell organelles. Several protozoa express photosynthetic pigments thus coloured. (Ex. *Euglena*) Many protozoa exhibit complex life cycles with multiple stages. (Ex. *Plasmodium*).

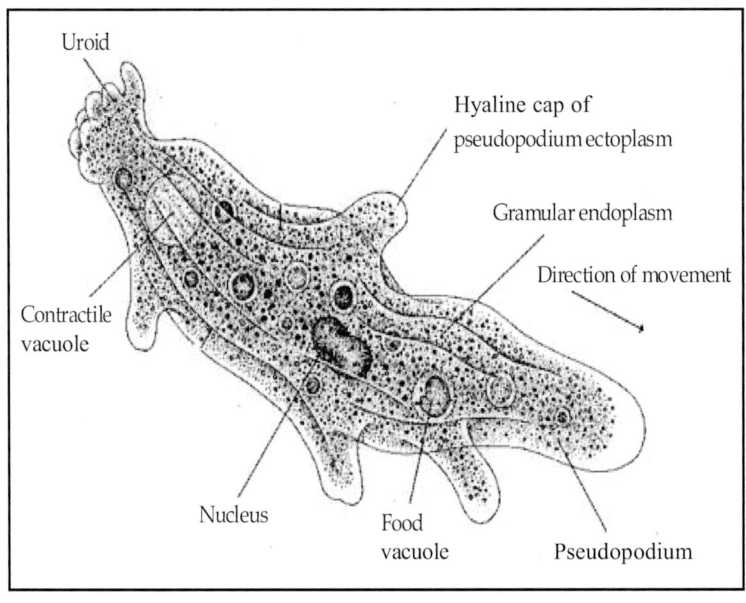


*Amoeba*

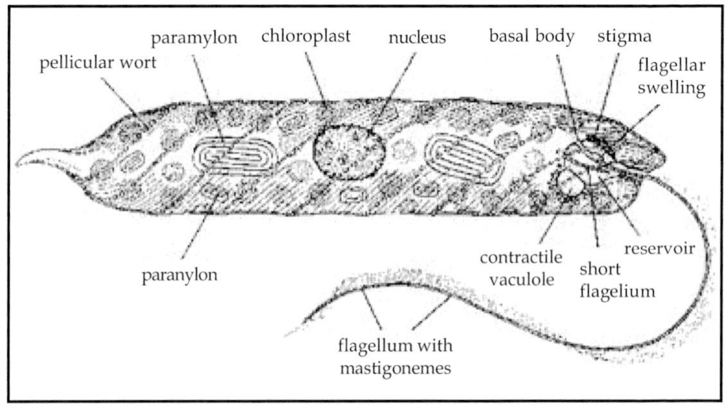


*Euglena*

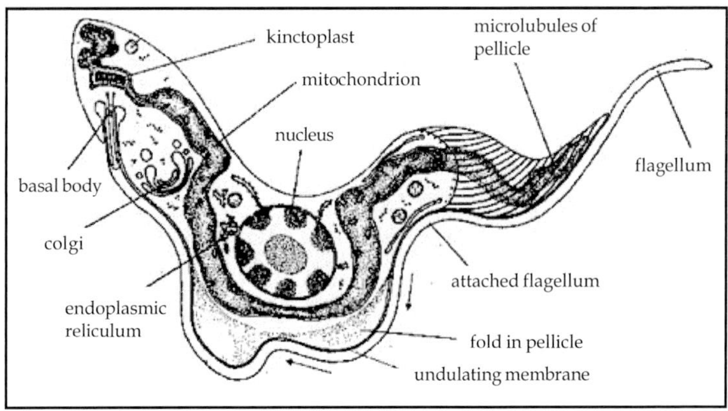


*Trypanosoma*

*Cyst:* Many protozoa form resistant cysts at certain times of their life cycle. They are able to survive in adverse environmental conditions such as desiccation, low nutrient supply and even anaerobiosis. In parasitic protozoa, the cyst stage is often used to transmit from one host to another.

## Locomotive Organs

The motility is the key character of protozoa and the mode of motility varies with organisms. The following are the possible ways by which the protozoa motility occurs.

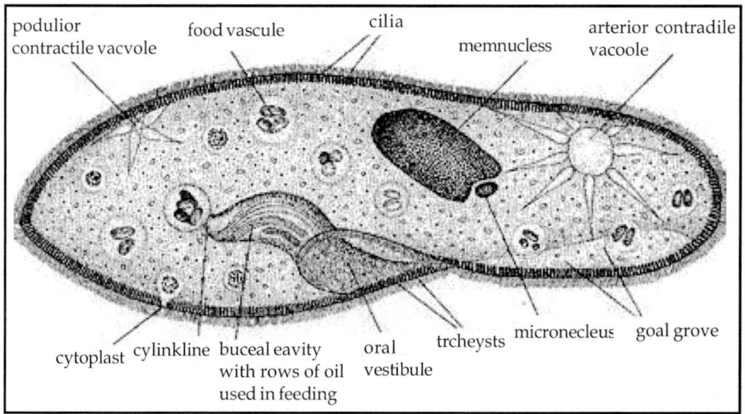


*Paramecium*

Fine structures of some common protozoa

1. *Pseudopodia (False foot) :* Temporary projection of part of cytoplasm of those protozoa which don't have rigid cell are referred as pseudopodia. Ex. *Amoeba.* The pseudopodia also helpful to capture food through engulfing.
2. *Flagella :* Fine filamentous extension of cell similar to that of bacteria. The number varies from 1 to 8. Ex. *Euglena.*
3. *Cilia :* Fine short thread like extension from cell used for locomotion and ingestion of food. Ex. *Paramecium*
4. *Gliding motility :* Ex. *Sporozoa.*

## Feeding

The ingestion is the primary mode of feeding habit of protozoa.

1. *Pseudopodia* of *amoeba* used to feed the prey by phagocytosis
2. *Cytosome or oral groove :* An opening of ciliate protozoa through which food is ingested. It ranges from simple round opening to slit like structures.

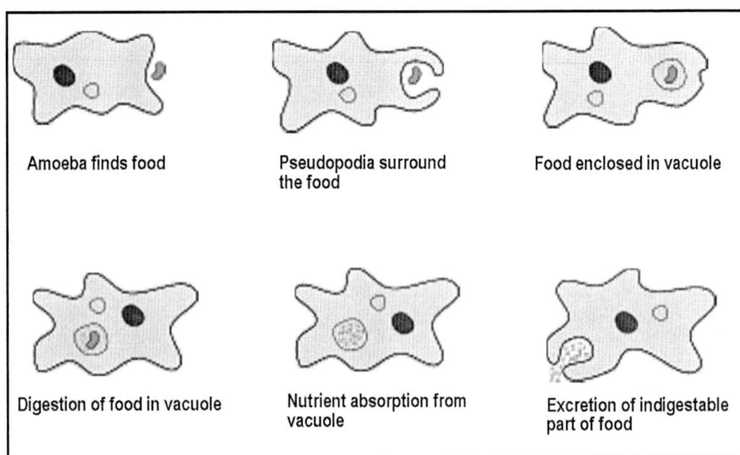


Feeding by Pseudopodla Ex. *Amoeba*

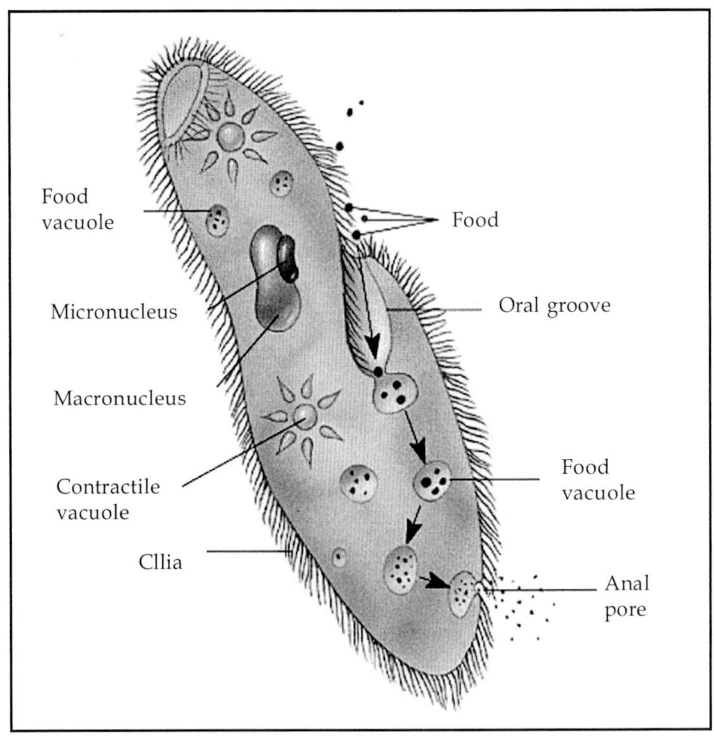


Feeding by oral groove (Cytosome) Ex. *Paramecium*

## Protective structures

Since the protozoa are habitat of aquatic nature, they are provided with protective structures. They are

1. *Mucocysts* : Certain ciliates secrete a mucilage substance from vesicles called mucocysts.
2. *Trichocysts* : Expulsion of harpoon like structures (spear like missile structures) which will attack the enemies.
3. *Toxicysts* : Thread like tubular structure of ciliate protozoa excrete toxins which used to paralyze and capture the prey.

## Reproduction

As a general rule, protozoa multiply by asexual reproduction and few have sexual reproduction too.

*Asexual reproduction*

1. *Binary fission* : Ex. *Paramecium, Amoeba*
2. *Multiple fission* : Single cell divided into many cells Ex. *Plasmodium*.
3. *Budding* : Ex. *Suctoria*

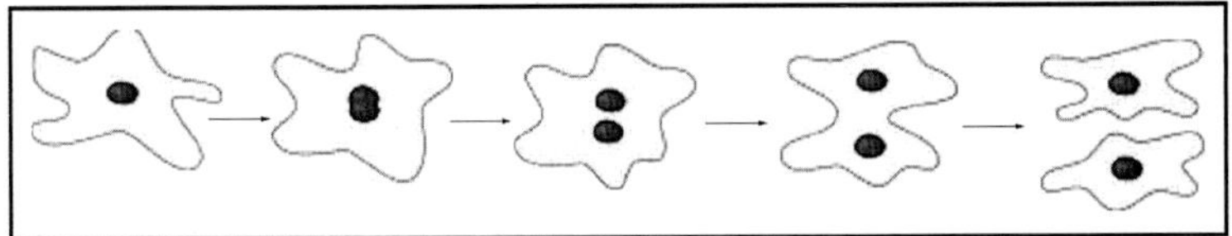

Binary fission of Amoeba

*Sexual reproduction*

1. *Sexual fusion of two gametes* : Ex. *Amoeba*
2. *Conjugation* : Transfer of genetic material from one organism to another through cilia Ex. *Paramecium* and *Euplotes*.

## Taxonomy

The protozoa are grouped into 5 different groups based on the morphological, ecological and evolutionary relationships. The following table shows the characters of individual groups.

| S. No. | Group | Common name | Example | Habitat | Common disease |
|---|---|---|---|---|---|
| 1. | Mastigophora | Flagellate | *Trypanosoma leishmania* | Fresh water, animal parasites | African sleeping sickness |
| 2. | Euglenoids | Phototrophic flagellates | *Euglena* | Fresh water and marine | - |
| 3. | Sarcodina | Amoebas | *Amoeba* | Animal parasites | Amoebiosis |
| 4. | Ciliophora | Ciliates | *Paramecium* | Animal parasites and rumens | Dysentery |
| 5. | Apicomplexa | Sporozoans | *Plasmodium* | Animal and insect parasites | Malaria |

*Economic importance of protozoa*

1. Protozoa act as link of food chain of aquatic and soil ecosystem.

   Phytoplankton → Zooplankton → Primary consumers → Secondary consumer. The protozoa are the zooplanktons which are referred as secondary producers.

2. Ecological balance of microbial communities in aquatic and wet land ecosystem was maintained by protozoa. They eat the prokaryotes so as to maintain their population in an ecosystem.

3. Protozoa play a major role in waste water treatment to remove the solid wastes (Ex. Anaerobic protozoa - *Metopus, Sporodinium* and aerobic protozoa – *Bodo, Paramecium*).

4. Diseases to human beings Ex. Malaria - *Plasmodium vivax; P. malariae* and *P. falciparum.;* African sleeping sickness - *Tryphanosoma leishmania;* Ambebiosis - *Amoeba* and *Entamoeba.*
5. Act as molecular tool to research on animal cell behaviors.

CHAPTER

9

# Virus

Virus can be defined as a genetic element containing either DNA or RNA that replicates in host cells as intracellular parasites but is characterized by having an extracellular state.

Virus can use the metabolic machinery of the host cell and can modify the genetics of the host cell. Virus has both extracellular and intracellular state.

## Extracellular State

Outside the host cell, the virus is a minute particle containing nucleic acid surrounded by protein, which is referred as virions or virus particles. Virions are inert and have no biosynthetic and metabolic functions. Extracellular state or virus particle is the perfect stage of virus to study the morphology and chemistry.

## Intracellular State

The active state in which the virus replicate in the host cell. When the virus nucleic acid is introduced to host cell and replication starts, the process is referred as infection. The cell in which infection of virus and replication takes place is referred as host cell.

## Chemistry of Virus

All the living cells have double strand DNA as their genetic material. In contrast viruses can either DNA or RNA as their genetic material and it can be either single strand or double strand. The third group of virus which contain RNA as genome, but replicate by DNA intermediate, used both DNA and RNA as their genetic material in their reproductive stage.

The viruses can be divided into following groups based on their nucleic acid and their form (single/double strand).

---

- DNA containing virus
  Single strand DNA virus
  Double strand DNA virus
- RNA containing virus
  Single strand RNA virus
  Double strand RNA virus
- RNA $\rightarrow$ DNA virus

---

Viruses are also divided into three major groups based on their host as 1. Plant viruses 2. Animal viruses and 3. Bacteriophages.

## Structure of Virions

The structure of virus varies to spherical, rod, complex structures and the size of virus is in nanometers (nm). The size ranges from 20 to 200 nm.

Ex. Small pox virus - 200 nm (smaller than smallest bacterium - *Nanobacterium*).

Polio virus - 28 nm (about the ribosome size).

The genome (genetic material in terms of base pairs of either DNA or RNA) size of each virus also varies.

- Herpes virus - ds DNA - 150 kbp (1 kbp = 1000 bp)
- Polio virus - ss RNA - 7 kbp.

- Cow pea mosaic virus - ss RNA - 9 kbp.
- Lambda (λ) Bacteriophage - ds DNA - 48 kbp.

The structure of virions is quite diversified. The nucleic acid of virion is always located within the particle and surrounded by protein coat called *capsid. (Protein coat / shell / capsid* are often used interchangeably).

The protein coat is always formed of a number of individual sub units referred as *capsomeres.* Capsomeres are arranged in a precise and highly repetitive pattern around the nucleic acid. These morphological sub-units can be seen under electron microscope.

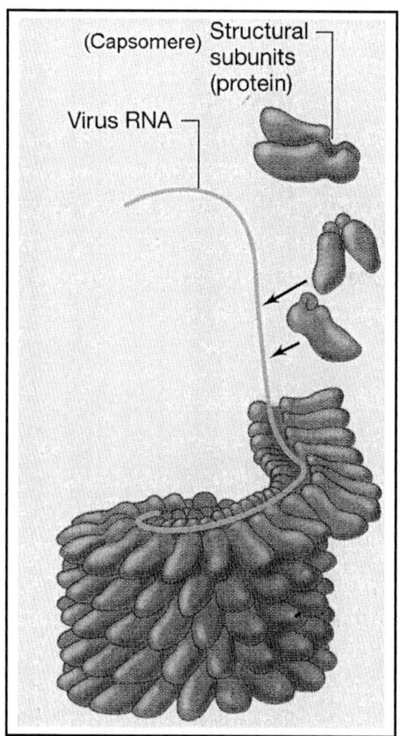


Fine structure of a virion Ex. Tobacco Mosaic Virus

The complete complex of nucleic acid and protein packed as virus particle is called as *nucleocapsids.*

Some viruses are covered with lipid bilayer which are referred as *envelopes.* Such viruses are referred as enveloped viruses and without envelopes as naked viruses.

## Virus Symmetry

Symmetry refers the way in which protein sub units are arranged in the virus coat. The rod shaped viruses are called as helical symmetry and the spherical shaped viruses are with icosahedral symmetry.

Ex. for helical symmetry - Tobacco mosaic virus; Icosahedral symmetry - adeno virus.

## Envelopes

Complex membrane structures surrounding the nucleocapsid are referred as envelopes. Enveloped viruses are common in animal viruses (HIV is an enveloped virus). Usually, the envelopes are of lipoprotein nature. The lipid portion is derived from host cell and the protein portion is synthesized by virus. They provide the shape of the virus.

## Complex Viruses

Some group of viruses have complex structures with several shapes. They provide head, tail, end plates and tail fibres. The maximum of 20 different proteins will be present in these viruses.

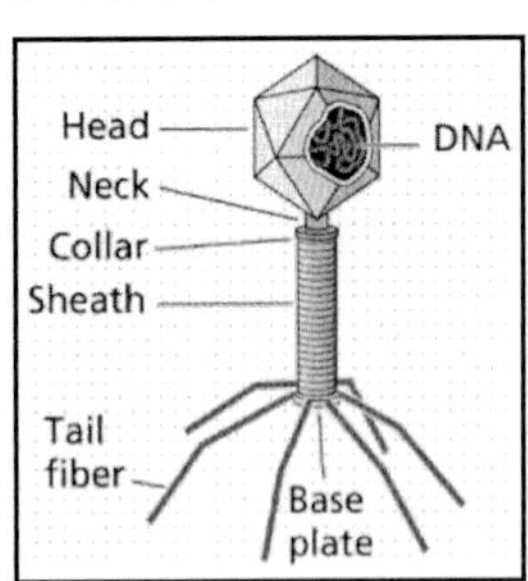


Structure of T2 Bacteriophage

## Viral Enzymes

As extracelluar particle, virions are metabolically inert. When infect the host cell, they can able to produce some enzymes which are essential for their replication.

1. *Nucleic acid polymerase :* Either DNA polymerase or RNA polymerase depends upon their genome.
2. *Reverse transcriptase :* Some RNA viruses (ex. HIV like retrovirus) have this enzyme to produce DNA from RNA.
3. *Neuraminadase :* To break down the glycosidic bonds of host cell to enter and to rapture the host cell.
4. *Lysozyme :* Especially bacteriophages produce lysozyme to make a hole in the bacterial cell wall for entry.

## Virus Replication

Since, virus doesn't have any system for replication by itself, it should induce the host cell to synthesize all essential components needed for complete virus. Then, these components must assemble into proper structure, and new virions must escape from the cell and infect the other cells. The sequence of virus replication is as follow:

1. *Attachment* (adsorption) of virion to a susceptible host cell
2. *Penetration* (injection) of virion or its nucleic acid into cell

*Early steps in replication :* during which the host cell biosynthetic machinery is altered by virion as a prelude to virus nucleic acid synthesis. Virus specific enzymes are typically made.

3. Replication of virus nucleic acid
4. Synthesis of protein sub-units
5. Assembly (Packing) and Escape or release

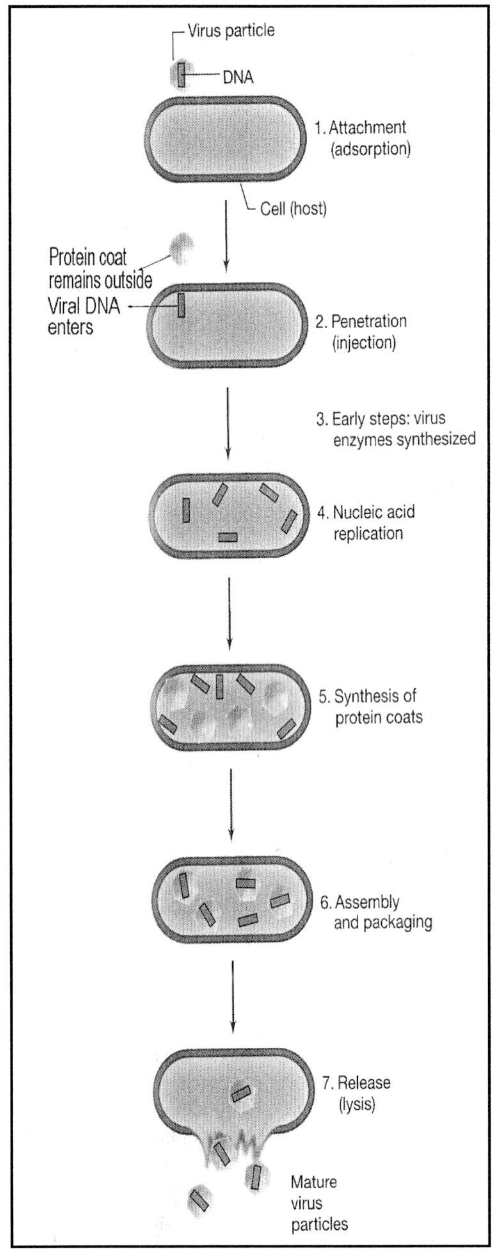


Virus Replication Cycle in Host Cell

## Attachment

Attachment of virus particle with host cell is mainly based on the specific interaction between host and virus. The virus has one or two specific proteins on the outside of the virion with specific activity. This protein will recognize some host cell surface components such as proteins, polysaccharides, lipoproteins etc. These compounds are referred as receptors. If the receptor is removed in the host cell, the attachment of virus is not possible on the host cell.

Ex. Some bacterial flagella or pili act as receptors; For influenza virus, glycoprotein of red blood cell acts as receptor; For T2 bacteriophage, core polysaccharide of outer membrane of *E. coli* acts as receptor. The resistant host will have no receptor for virus or the receptor is protected by some physical means (Ex. Capsules of bacteria give resistance by blocking the receptors).

## Penetration

The penetration of virus into the host cell is mainly based on the surface morphology of the host cell. The penetration varies from host to host (plant and bacteria have cell wall and animals don't have cell wall).

The bacteriophages (T2 phage) first attached its tail fibres with receptor (core protein); By means of lysozyme, the phage made a hole in the cell wall and injected the nucleic acid by the help of tail. The animal viruses as a whole virion penetrate into the host cell through phagocytosis. Plant viruses enter into host cell by mechanical injuries or by some insect's action.

## Virus Restriction & Modification by Host

When the viral nucleic acid enters to cytoplasm of the host cell, as a protective mechanism host cell can able to digest the nucleic acid by means of restriction endonuclease (for DNA) or by RNAse (in case of RNA). To avoid that, the viral nucleic acid or genome modify its base pairs so that the restriction enzymes cannot react on it. By addition of glucose (glucosylation) or by addition of methyl group (methylation), the modification of viral genome is possible.

## Replication of Viral Nucleic Acid and Proteins

New viral proteins and nucleic acids should be synthesized with the help of host DNA. The genetic system of host is used by the viral nucleic acid to produce capsid and nucleic acid. Nucleic acids were synthesized by means of DNA or RNA replication using the enzymes DNA polymerase or RNA polymerase respectively. The protein part of virions is synthesized as protein synthesis. The genetic informations of virus (DNA or RNA) should be converted into mRNA (called as transcription). The mRNA synthesis shows diversity among the viruses. Following diagram shows how the mRNA synthesis occurs from different viral genomes namely ssDNA, dsDNA, ssRNA and dsRNA.

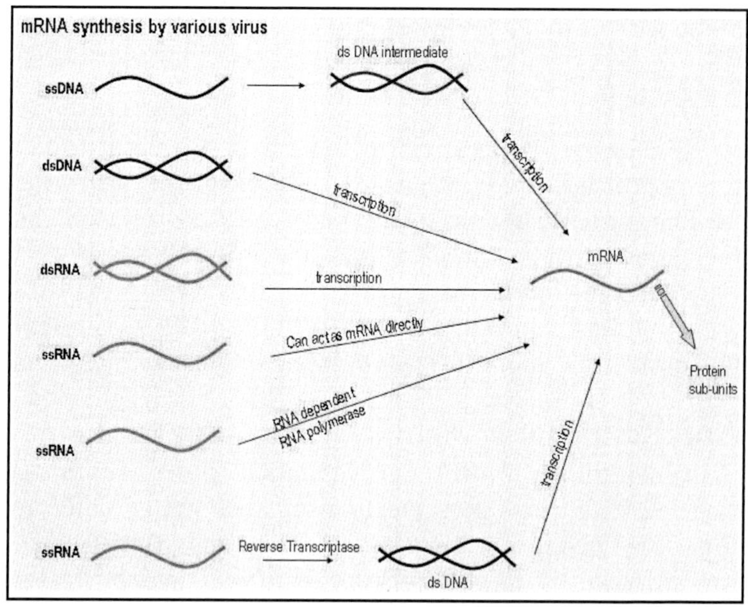

After mRNA synthesized, viral protein sub-units are synthesized. The early proteins - synthesized soon after infection (enzymes) and late proteins - synthesized later part (coat protein).

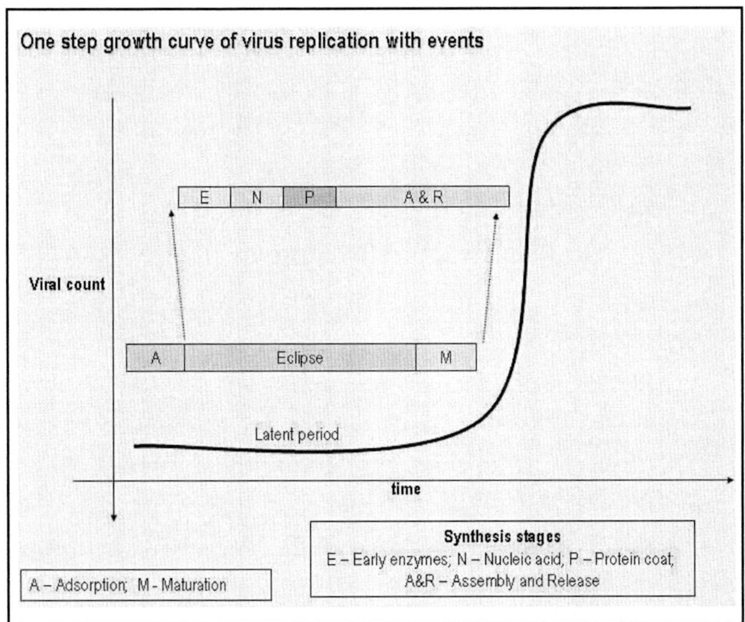


The latent period refers the period which the presence of virus in the host cell cannot be detected. During that period, the synthesis of enzymes, nucleic acid and protein synthesis take place.

After the protein coat and nucleic acid synthesis, the assembly takes place leads to completion of viral particle which will lyse the host cell; escape and infect the next host cell. Above diagram shows the one step growth curve (means one lytic cycle) of virus against time. The population of virus is expressed as *plaque forming units.*

These types of viruses, which kill the host cell during release, are referred as *lytic viruses* and the cycle is referred as *lytic cycle.* But some bacteriophages will multiply inside the host cell and release to the environment after assembly without killing the host cell are referred as *temperate phages* and the cycle is referred as *lysogenic cycle.*

*1. Bacteriophages*

The viruses which attack the bacteria and multiply are referred as bacteriophages. Most of the bacteriophages studied so far are belonged to the virus affecting *E. coli* and *Salmonella.*

- Most of the bacteriophages are naked
- Most of them structurally complex nature
- Head, tail, tail fibres, base plates are some parts of these viruses
- Most of these viruses are used as molecular tool for genetic engineering
- The shape, size, chemistry and replication method are diversified.
- ssRNA, dsRNA, ssDNA and dsDNA are as genomes.

*Replication methods of bacteriophages*

1. *MS2 :* This single strand RNA phage uses its RNA as mRNA and protein synthesized. The complimentary RNA also synthesized from which viral RNAs synthesized.
2. *$\phi$X174 :* This ssDNA phage has single strand *circular* DNA which duplicate by rolling circle replication.
3. *M13 :* This ssDNA phage has *linear* DNA. The DNA synthesized in cytoplasm and protein coats are synthesized and accumulated in the cytoplasmic membrane. The linear DNA when released from host covered by protein coat and released. This is a temperate phage which will not kill the host cell during replication. (A slight decline in the growth of

bacteria will occur). The cycle is referred as *lysogenic cycle* and the phage is *temperate phage* and host bacterium is *lysogen.*

Based on the nucleic acid, strands, shapes, the bacteriophages are grouped as follows:

| S. No. | Group | Example | shape |
|---|---|---|---|
| 1. | ss RNA phages | MS2 |  |
| 2. | ds RNA phages | ϕ6 |  |
| 3. | ss DNA phages | ϕX174 | 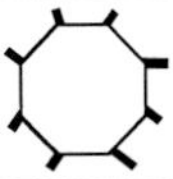 |
| | | M13 |  |
| 4. | ds DNA phages | T3, T7<br>Lytic phage | 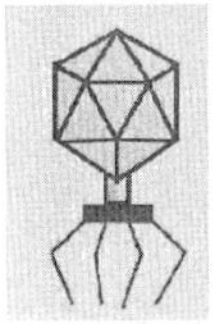 |
| | | Mu<br>Mutagenic phage | 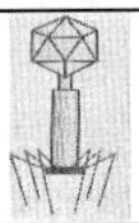 |
| | | Lambda<br>Temperate phage |  |
| | | T2, T4<br>Temperate phage | 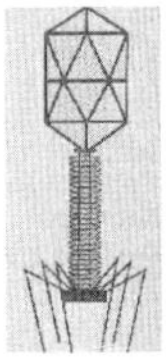 |

4.*T2,T4, T3,T7 phages* : They are ds DNA phages which are lytic phages and the cycle is lytic cycle. (25 minutes per cycle). The lytic cycle of these phages occurs as discussed earlier.

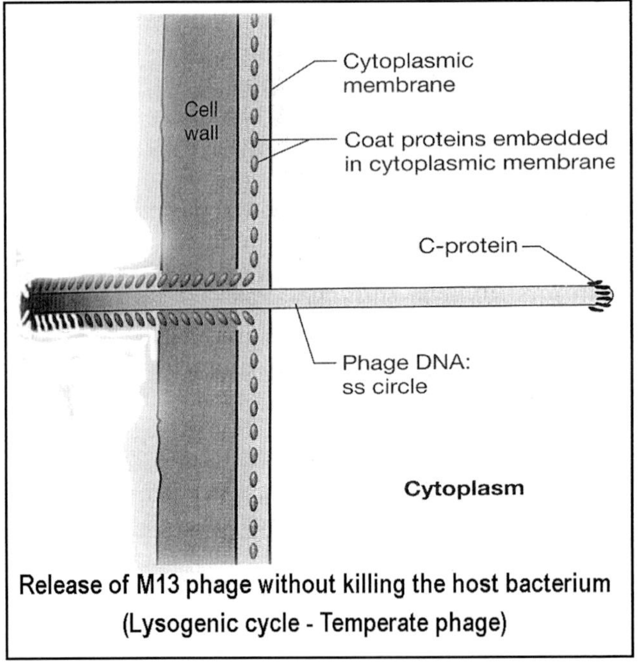


**Release of M13 phage without killing the host bacterium (Lysogenic cycle - Temperate phage)**

5. *Lambda phage* : This dsDNA phage is a temperate phage. After penetration, the lambda phage DNA will integrate with the bacterial DNA and will persist for ever. This state of phage is referred as *prophage*. When the prophage is induced (either by chemical or physical means) it become lytic phage and replicate in the host cell and by killing the host the virions will escape.

6. *Mu Phage* : This dsDNA phage is also a temperate phage like lambda phage. It has the unusual property like transposable elements (which can move from one portion of genomic DNA to another part) and thereby cause mutation. They are referred as biological mutagens.

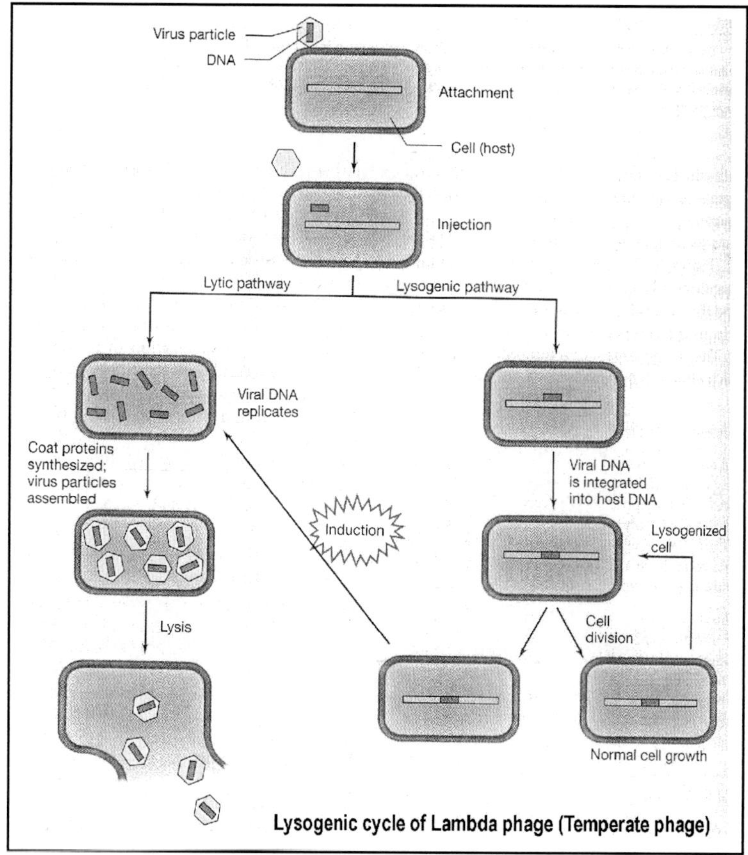

**Lysogenic cycle of Lambda phage (Temperate phage)**

2. *Animal Viruses*

- The animal viruses are both DNA and RNA viruses. Retro virus is a RNA virus but can have DNA as an intermediate (referred as RNA → DNA virus).
- The animal viruses enter into host cell as whole virion through phagocytosis.
- The viral nucleic acid will be produced at the nucleus of host cell and the protein coat will be produced at cytoplasm and assembly takes place.
- Most of the animal viruses are enveloped.

- The animal viruses vary their infection pattern on their host.

## Types of Infection

- *lytic infection* : destroys host cells.
- *persistent infection* : host cell continues to shed virus over long time. Cell gradually becomes recognizably poorer (recognized as cytopathic effect, or CPE), eventually "crumps out".
- *Oncogenes* : Infection by certain viruses causes cells to change, become cancerous. Responsible genes are called *oncogenes* (tumor-producing genes). Viral oncogenes have also been found in uninfected cells. These are genes involved in regulation of cell cycle; when defective, normal regulatory control is lost and cell can become cancerous.
- *Persistent infection* : The viruses released slowly from the host cell without affecting the host cell.
- *latent infection* : virus genes may not be expressed for long time (ex. many Herpes infections). Not the same as lysogeny – genes are not integrated into host chromosome.

The animal viruses are grouped in to many families based on nucleic acids, envelopes, etc.

## Retrovirus Replication

The structure of Retrovirus (Human Immunodeficiency Virus) which has ssRNA as its genome and surrounded by core protein and envelope.

The retrovirus is the RNA containing enveloped virus, has the specialized system of replication in the host cells. The virus temporarily synthesizes DNA by means of *reverse transcriptase* enzyme and the DNA get integrated into

| S. No. | Groups | Enveloped / naked | DNA / RNA | Imp. example |
|---|---|---|---|---|
| 1. | Parvovirus | N | DNA | Mice minute virus |
| 2. | Papovirus | N | DNA | |
| 3. | Adenovirus | N | DNA | |
| 4. | Iridovirus | N | DNA | |
| 5. | Hepadnavirus | E | DNA | Hepatitis B virus |
| 6. | Pox virus | E | DNA | Small pox virus |
| 7. | Herpes virus | E | DNA | Chicken pox |
| 8. | Picomavirus | N | RNA | Polio virus, Hepatitis A virus |
| 9. | Reovirus | N | RNA | Blue tongue virus |
| 10. | Togavirus | E | RNA | |
| 11. | Rhabdovirus | E | RNA | Rabies virus |
| 12. | Orthomyxovirus | E | RNA | Influenza virus |
| 13. | Coronavirus | E | RNA | SARS virus |
| 14. | Bunyavirus | E | RNA | |
| 15. | Retrovirus | E | RNA | HIV |
| 16. | Paramyxovirus | E | RNA | Measles virus |
| 17. | Arenovirus | E | RNA | |

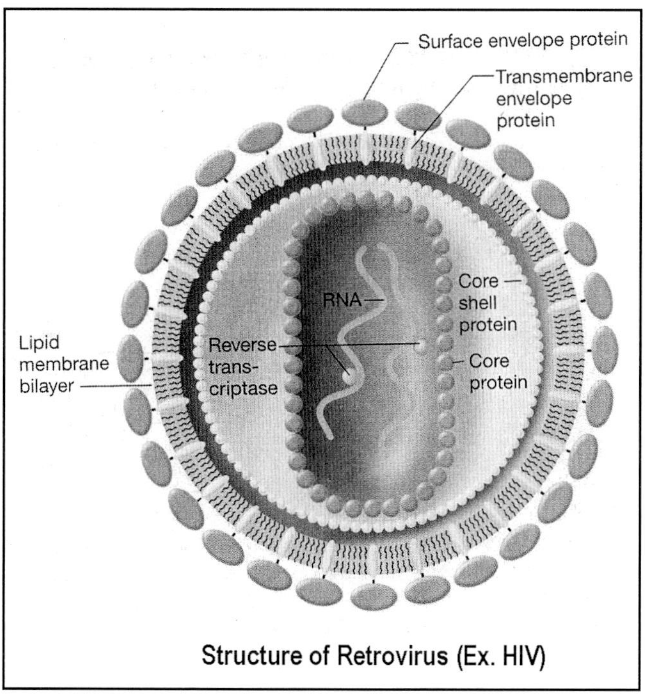


Structure of Retrovirus (Ex. HIV)

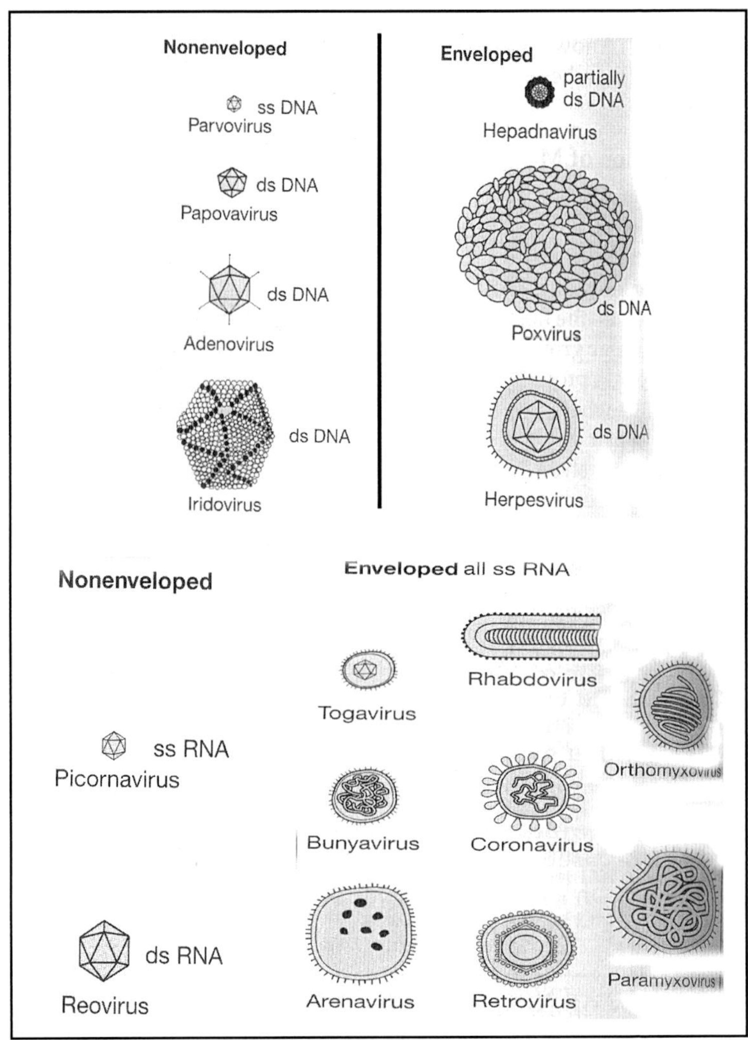


Some animal viruses

chromosome of host DNA. After some period, the DNA integrated in the genome synthesize the RNA and protein coat. After assembly, while escape from the cell, the cytoplasmic membrane of host cell provides the envelope and the virion escaped the host by lysis.

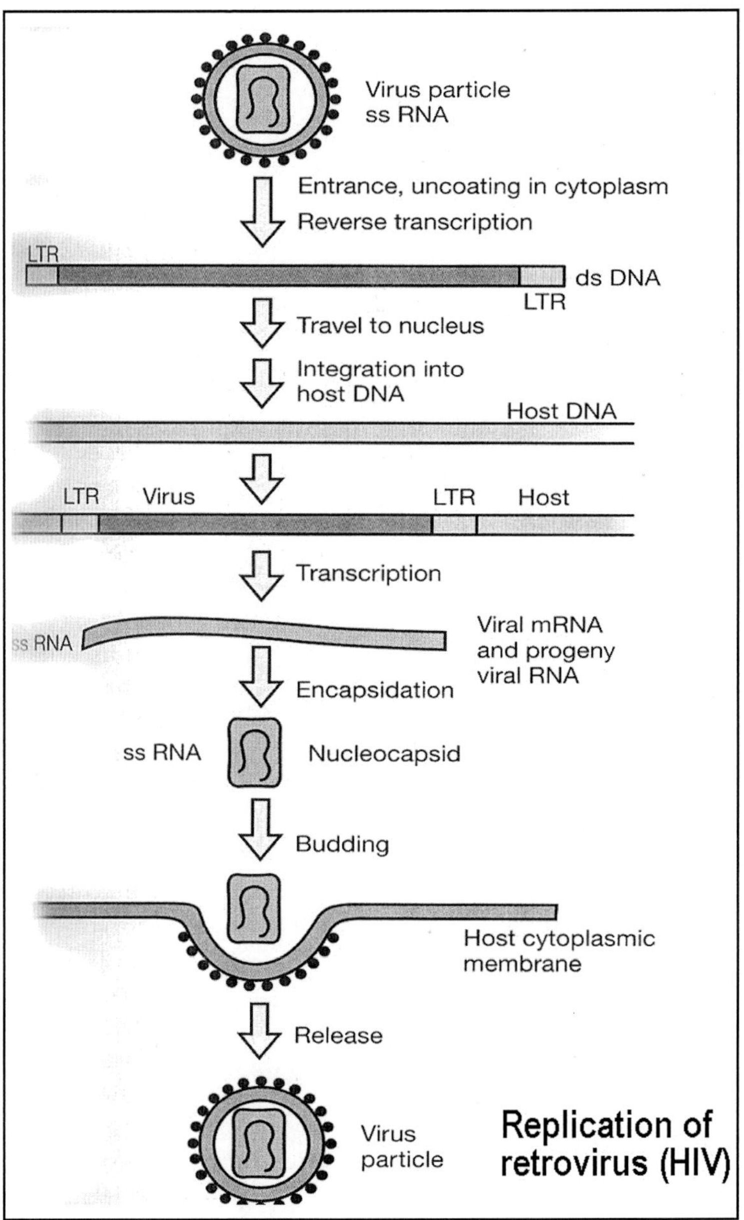


Replication of retrovirus (HIV)

*3. Plant Viruses*

- Most of the plant viruses are naked viruses.
- They don't have specific mechanism for entering the host plants. The injuries, insect vectors are the means of entry.
- The cell wall and cuticle layers of plant are the barrier for the entry of virus.
- The virus diseases on plants are very rare because infection is not strong enough to kill the plant.
- Mostly the virus infection can be recognized by mosaic like leaf pattern, in which the loss of chlorophyll leads to yellow in colour.
- Some viruses multiply in the host plant without showing the symptom referred as latent infection.

There are four important groups of plant viruses.

| **S. No.** | **Virus group** | **Common virus** |
|---|---|---|
| 1. | Single strand RNA virus | Tobacco mosaic virus<br>Turnip crinkle virus<br>Cucumber mosaic virus |
| 2. | Double strand RNA virus | Wound tumor virus |
| 3. | Single strand DNA virus They are referred as Gemini virus (Usually found as pair) | Maize streak virus<br>Tomato golden mosaic virus |
| 4. | Double strand DNA virus | Cauliflower Mosaic virus (CaMV) (This virus can modify the plant genome. So used as molecular tool in the field of plant biotechnology |

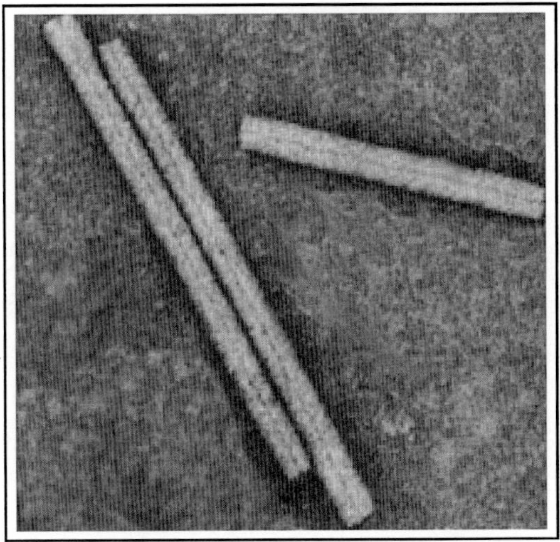

EM of Tobacco Mosaic virus

*4. Viroids*

Viroids are the smallest infectious agents which are having a circular RNA alone as genome (they don't have even protein coat too). They use DNA dependent RNA polymerase of host plant for their multiplication. The viroids infect the plants and cause diseases. Ex. Potato spindle tuber viroid. (It consists of 359 ribonucleotides and is characterized by numerous intra molecular base-pairings that lend stability to the structure).

*5. Prions*

Prions are infectious protein particles that cause several diseases in animals and humans. Prions are the only infectious agents that do not contain any nucleic acids.

Prion diseases are often called spongiform encephalopathies because of the post mortem appearance of the brain with large vacuoles. They cause brain diseases in sheep (scrapie disease), cows ("mad cow disease"), humans (Kuru and Creutzfeld-Jakob disease) and other animals.

## Creutzfeldt-Jakob Disease

- occurs worldwide and usually becomes evident as dementia.
- *Very rare:* affects one person in a million. 10 to 15 percent of cases are inherited.
- Typically begins in 60's with a loss of memory.
- Over several weeks the mental deterioration progresses to dementia, abnormalities of vision or coordination, rigidity, and involuntary.
- Death usually occurs within six months, and at necropsy the brain shows a "spongy" texture.

## Mad Cow Disease

- Cows become uncoordinated and unusually apprehensive.
- Disease traced to use of sheep tissue as food supplement for cows over long period of time in United Kingdom. Practice was begun in late 1970s.

CHAPTER 10

# Microbial Metabolism

The essential nutrients and their function in microbial cells

| Element | % of dry weight | Source | Some Laboratory reagents used in media preparation | Function |
|---|---|---|---|---|
| **Major elements** | | | | |
| Carbon | 50 | organic compounds or $CO_2$ | Glucose, sucrose, lactose, malic acid, mannitol, starch, dextrin, | Main constituent of cellular material |
| Oxygen | 20 | $H_2O$, organic compounds, $CO_2$, and $O_2$ | Water, air and organic compounds | Constituent of cell material and cell water; $O_2$ is electron acceptor in aerobic respiration |
| Hydrogen | 8 | $H_2O$, organic compounds, $H_2$ | Water, air and organic compounds | Main constituent of organic compounds and cell water |
| Nitrogen | 14 | $NH_3$, $NO_3$, organic compounds, $N_2$ | Ammonium sulphate, sodium sulphate, sodium nitrate, sodium nitrite, peptone, urea, tryptone, soybean extract, beef extract, yeast extract | Constituent of amino acids, nucleic acids nucleotides, and coenzymes |

*Contd...*

| | | | | |
|---|---|---|---|---|
| Phosphorus | 3 | inorganic phosphates ($PO_4$) | $KH_2PO_4$, $K_2HPO_4$, $NaH_2PO_4$, $Na_2HPO_4$ | Constituent of nucleic acids, nucleotides, phospholipids, LPS, teichoic acids |
| Sulfur | 1 | $SO_4$, $H_2S$, So, organic sulfur compounds | $Na_2SO_4$, $MgSO_4$, $Na_2S_2O_3$, cystine like aminoacids | Constituent of cysteine, methionine, glutathione, several coenzymes |
| Potassium | 1 | Potassium salts | KCl, $KH_2PO_4$, $K_2HPO_4$, | Main cellular inorganic cation and cofactor for certain enzymes |
| Magnesium | 0.5 | Magnesium salts | $MgSO_4$, MgCl, | Inorganic cellular cation, cofactor for certain enzymatic reactions |
| Calcium | 0.5 | Calcium salts | $CaCl_2$ | Inorganic cellular cation, cofactor for certain enzymes and a component of endospores |
| Iron | 0.2 | Iron salts | $FeSO_4$, Fe – EDTA, $FeCl_3$, Fe citrate | Component of cytochromes and certain nonheme iron-proteins and a cofactor for some enzymatic reactions |
| **Micronutrients** | | | | |
| Cobalt | Trace | Inorganic salts | Cobalt chloride | Vitamin B12, cofactor |
| Copper | Trace | Inorganic salts | Copper sulphate | Respiration, ETC, cofactor |
| Manganese | Trace | Inorganic salts | $MnCl_2$ | Activator of many enzymes |
| Molybdenum | Trace | Inorganic salts | Sodium molybdate | Flavin containing enzymes, composition in important enzymes |
| Zinc | Trace | Inorganic salts | Zinc chloride, zinc sulphate | DNA, RNA polymerase enzymes, DNA binding proteins |

Apart from these macro and micro nutrients, the microorganisms may require a small amount of certain organic compounds for growth, because of their essentiality and unable to synthesis from available nutrients.

Common vitamins required in the nutrition of certain bacteria

| Vitamin | Coenzyme form | Function |
|---|---|---|
| p-Aminobenzoic acid (PABA) | - | Precursor for the biosynthesis of folic acid |
| Folic acid | Tetrahydrofolate | Transfer of one-carbon unit and required for synthesis of thymine, purine bases, serine, methionine and pantothenate |
| Biotin | Biotin | Biosynthetic reactions that require $CO_2$ fixation |
| Lipoic acid | Lipoamide | Transfer of acyl groups in oxidation of keto acids |
| Mercaptoethane-sulfonic acid | Coenzyme M | $CH_4$ production by methanogens |
| Nicotinic acid | NAD (nicotinamide adenine dinucleotide) and NADP | Electron carrier in dehydrogenation reactions |
| Pantothenic acid | Coenzyme A and the Acyl Carrier Protein (ACP) | Oxidation of keto acids and acyl group carriers in metabolism |
| Pyridoxine (B6) | Pyridoxal phosphate | Transamination, deamination, decarboxylation and racemation of amino acids |
| Riboflavin (B2) | FMN (flavin mononucleotide) and FAD (flavin adenine dinucleotide) | Oxidoreduction reactions |
| Thiamine (B1) | Thiamine pyrophosphate (TPP) | Decarboxylation of keto acids and transaminase reactions |
| Vitamin B12 | Cobalamine coupled to adenine nucleoside | Transfer of methyl groups |
| Vitamin K | Quinones and napthoquinones | Electron transport processes |

## Microbial Metabolism - Basis

Before discussing the microbial metabolism, it is essential to understand the energy flow, basic principles of energetics of living cells etc.

*Energy*

Defined as the ability to do work. The energy can be referred as Kcal or K Joules. One k cal is the energy required to raise 1°C of 1 kg of water. 1 calorie = 4.184 Joules

Chemical energy - referred as the energy released during oxidation of organic or inorganic compounds.

*First Law of Thermodynamics*

" Energy can neither be created nor be destroyed in the universe"

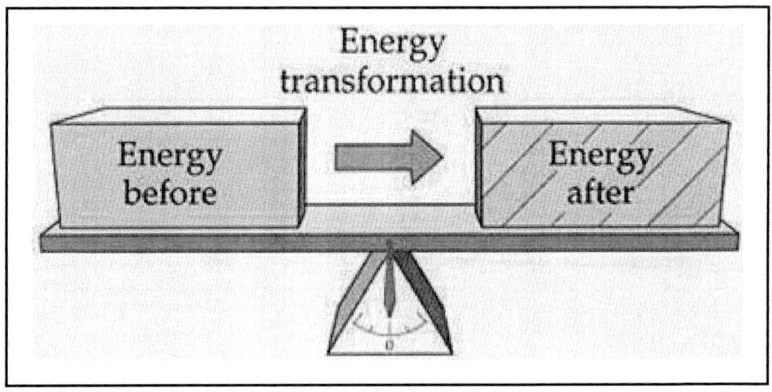


So, the living things should collect the energy from existing sources and convert them to a suitable form for the biological process.

Plants grab the energy from light and convert them to high energy chemical compounds. Similarly, animals can derive their energy by oxidizing the chemicals. Microorganisms especially the bacteria have both the ways (from chemicals and sun light)

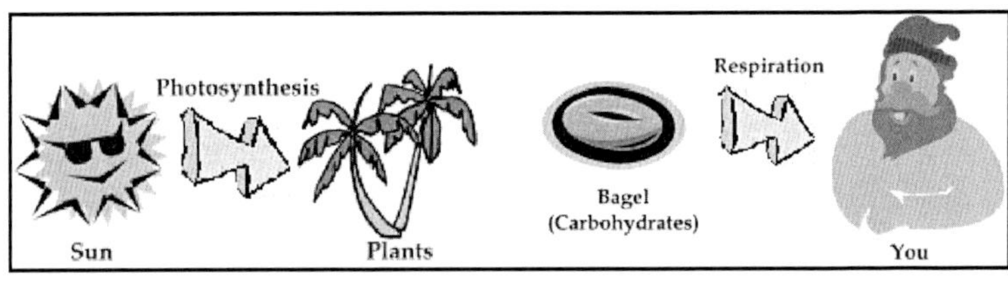

*Second law of Thermodynamics*

"In all the reactions, some of the energy involved loses its ability to do work"

We can refer this energy lose as "disorder". Generally living systems are "ordered", but universe always prefers disordered. The living systems are always against the disorder and the living systems have constant battle against disorder and when they lose the battle, they die.

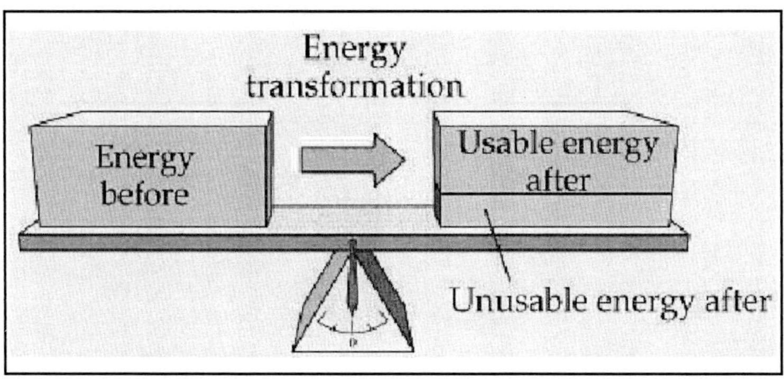


*Free energy (ΔG)*

The quantity of energy which is unable to do work in a reaction is referred as *entropy*. The energy released from a reaction , that is available to do some useful work is referred as *free energy*.

Total energy = free energy (ÄG) + entropy

A reaction may *release* or *require* free energy. If a reaction needs energy, it is referred as *endergonic reaction* and if a reaction releases energy, it is referred as *exergonic reaction*.

$A + B \rightarrow C + D$ + energy [exergonic reaction]

$A + B$ + energy $\rightarrow C + D$ [endergonic reaction]

ÄG of some compounds are

$H_20$ = -237.2; $CO_2$ = -396.4; $H_2$ = 0; $O_2$ = 0;

$N_2O$ = +104.2; Glucose = -917.3

*Negative ÄG is good and available to do work in the cell.*

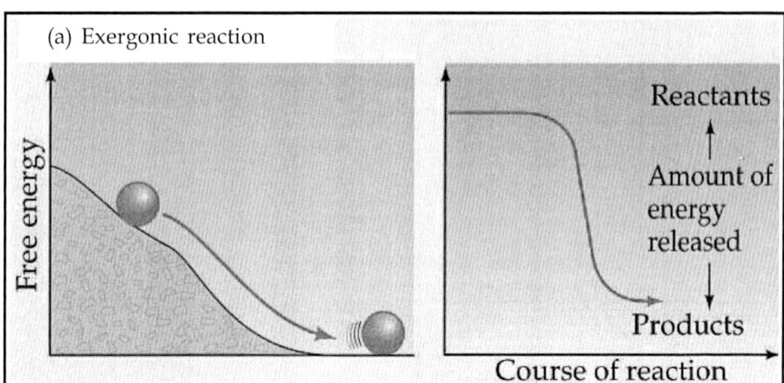


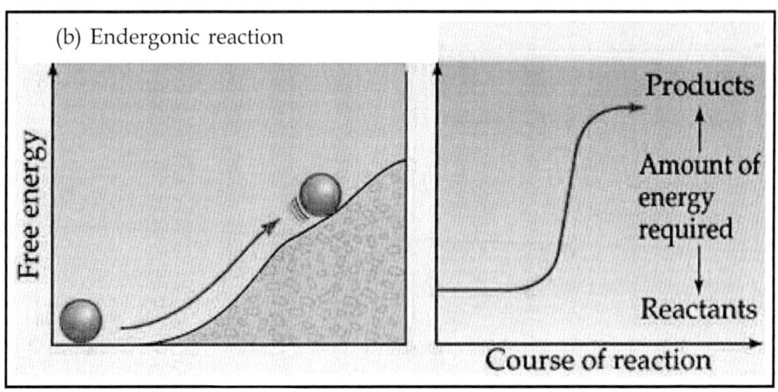


*Activation energy*

Referred as the energy required to bring all the molecules to reactive stage.

When the $H_2$ and $O_2$ gases kept together in a container will not react each other even after many years and need some energy to make them as reactive molecules to react and gives $H_2O$. That energy is referred as activation energy.

*Catalyst* is a substance to lower the activation energy requirement of a reaction and also to increase the reaction rate.

The *diagram below showing the lower down of activation energy (Ea) by means of catalytes*

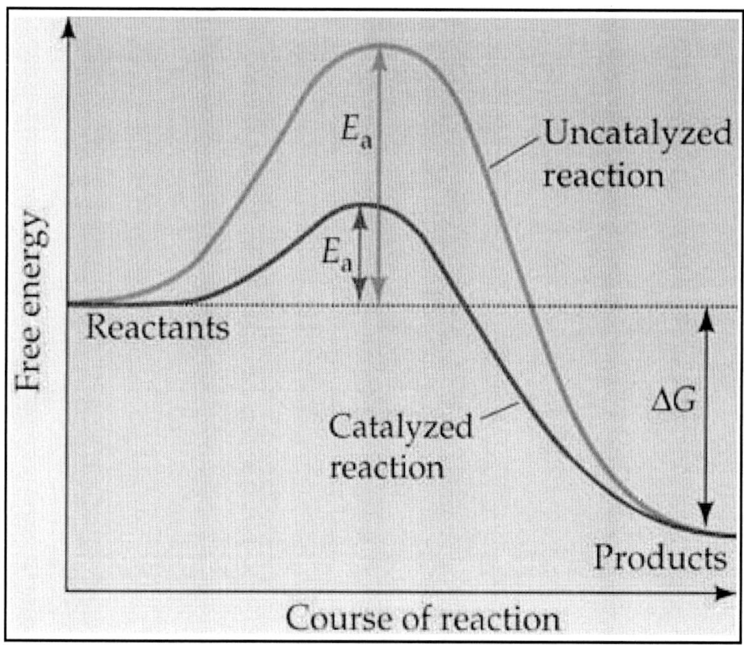


In biological systems, proteins involved in the catalytic activity are referred as *biocatalysts* or *enzymes*.

They increase the reaction rate of $10^8$ to $10^{20}$ times.

## Enzymes

They are protein in nature with the size of 1x $10^4$ to 1 x $10^6$ Dalton.

Normally some non-proteinecious low molecular substances were joint together with protein part to serve as enzymes.

Protein + Non-protein compounds → enzyme

The protein part is *Apoenzyme*; The non-protein is referred as *Co-enzyme* or *Prosthetic group*.

Both (apoenzyme and coenzyme/Prosthetic group) are referred as *Holoenzymes*.

The difference between co-enzyme and prosthetic group is that co-enzyme will have loose bound with apoenzyme (temporarily) whereas the prosthetic group will have tight bound with apoenzymes (permanently).

Ex. Co-enzymes - NADP, NAD, FAD, FAM

Ex. Prosthetic group - Heme group of cytochromes

Apart from the co-enzymes and prosthetic groups, the inorganic ions like Ca, Mg, Fe, Mn are required for the activation of enzyme and they are referred as *co-factors*.

Enzymes are very specific, efficient and can be inhibited by products.

$$A \xrightarrow{\text{Enzymes}} C + D$$

In this reaction, A is the substrate and C and D are the products. For this reaction, the substrate A should attach with the enzyme at specific site referred as *substrate binding site*. Then, they form a *substrate - enzyme complex* and lead to act on A to give C and D products. Then, C and D were detached from the enzyme and the enzyme return to its original position.

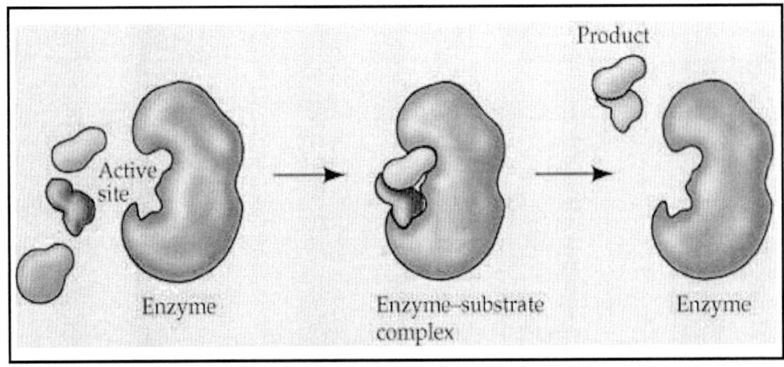


Function model of an enzyme

Microbial cells are as bags of enzymes. Each enzyme is located at appropriate place and active form. One enzyme converts a substrate to a product and that product may be served as substrate to another enzyme to form product –2. Like this, series of reactions will form a path way which is referred as *biochemical path ways* or *metabolic path ways*.

During this chain of reactions, some reaction may be *exergonic* and some may be *endergonic*. So, there is a chance for release of energy or chance for utilization of energy in these reactions.

The released energy from the exergonic reactions were stored in the form of high energy phosphate bonds in ATP (Adenosine triphosphates).

Similarly, whenever the endergonic reactions occur, the ATP break down its high energy bonds and thus released energy can be utilized.

So, the ATP can be referred as *energy carrier* in the biochemical pathways.

## Oxidation - Reduction reaction

In universe, oxidation - reduction reaction is the common method to extract the energy by living cells.

- All the molecules contain electron as part of atoms.
- These electrons can be disturbed.
- Each molecule can able to donate or accept the electron.

If a molecule or compound donates or loses its electrons, the reaction is referred as *oxidation* and if a molecule or compound receives or accepts electron, the reaction is referred as *reduction*.

*Note :* Remember the phrase "LEO says GER" LEO - Loss of Electron is Oxidation; GER - Gain of Electron is Reduction

For example,

If $H_2$ gas loses its electron,

$H_2$ ----------------→ $e^- + H^+$ will be the reaction.

The $H_2$ gas will give electron and proton ($H^+$) as the products. In this reaction, the $H_2$ loses its electron and hence referred as reduction. But, the reaction is not yet completed. If $O_2$ is present, it will accept the electrons and form water as product.

$2O_2 + e^- + H^+$ ----------------→ $2H_2O$.

Since, the oxygen received or accepted the electrons, the reaction is referred as reduction. Here, the hydrogen donates or loses the electron and hence referred as *electron donor* and the oxygen receives or accepts the electron and hence referred as *electron acceptor*.

Substrates or compounds vary their ability to donate or accept the electrons (in other words, to become oxidized or reduced). The tendency of a substrate to donate or accept the electron is referred as *oxidation – reduction potential or redox potential*. The redox potential of each substance was calculated by means of electrical volt with reference substance as $H_2$.

The redox potential is –ve means, the substance is rich of electrons, (means reduced form) and +ve means it already lost the electrons (means oxidized form). The reduced forms can give or donate electrons and oxidized form can accept the electrons.

If series of oxidation – reduction reactions occur, the electrons from one reaction can be carried out to another by means of electron carriers. They can accept the electrons from a reaction and can donate the same in some other reaction. They are called as *electron carriers*. In biological systems, many such carriers are available. Ex. NAD, NADP, FAD, FAM etc.

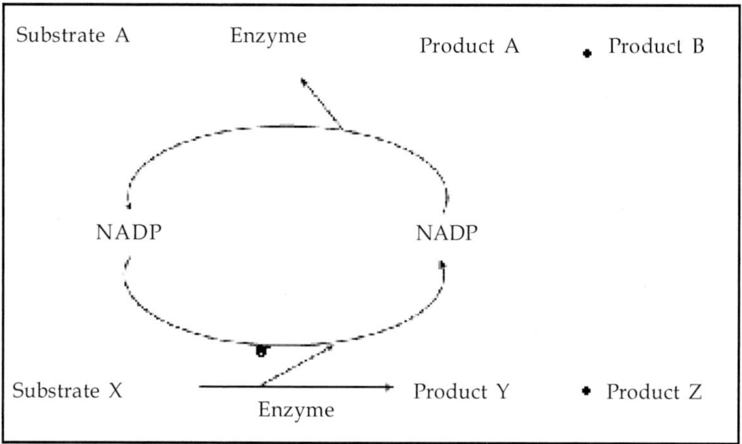


In the above reaction, when substrate X get oxidized, the electron released can be taken up by NADP (oxidized form) and reduced to NADPH. The NADPH (reduced form) then moves to some other reaction and can donate the electron to substrate A which may give the products A and B.

This is how the electron carriers functioned in oxidation-reduction reactions or in metabolic path ways.

## Phosphorylation

Addition of phosphorus is referred as phosphorylation. The energy released during oxidation reaction is stored in the form of *high energy phosphate bonds* in ATP. When energy is released, the ADP and Pi reacted and form ATP. This process of addition of Pi to from ATP is referred as phosphorylation.

*Note:* The high energy phosphate bonds are differ from normal phosphate bonds and the energy released from ATP is higher than glucose phosphates.

The phosphorylation or ATP formation occurs in three different ways in bacteria.

a. *Substrate level phosphorylation :* Energy released during oxidation of one organic molecule to another will be used for ATP production. There won't be any external electron acceptor and the organic substrate itself acts as electron acceptor. This kind of phosphorylation is referred as *substrate level phosphorylation.*

b. *Oxidative or Electron transport phosphorylation* : ATP formation during electron transport chain by proton motive force is referred as oxidative phosphorylation. The electrons are carried by several electron carriers and finally the terminal electron acceptor (like $O_2$) will accept the electron and form $H_2O$. The proton motive force of ATPase used for ATP synthesis.

c. *Photophosphorylation* : The ATP formation using light energy is referred as photophosphorylation. The proton motive force during photosynthesis is used for ATP synthesis.

## Microbial Metabolism

Metabolism refers the sum of biochemical reactions required for energy generation and the use of energy to synthesize cellular materials.

The energy generation component is referred as *catabolism* and the build up of macromolecules and cell organelles are referred as *anabolism.*

During catabolism, the energy is changed from one compound to another and finally conserved as high energy bonds of ATP. ATP is the universal currency for energy. When energy is required for anabolism, it may be sent as high energy bonds of ATP which has the value of 8 kcal per mole.

Based on the source of carbon, the microbes can be divided into two groups namely, autotrophs and heterotrophs. *Autotrophs* utilize $CO_2$ as sole carbon source and *heterotrophs* use organic carbon as sole carbon source.

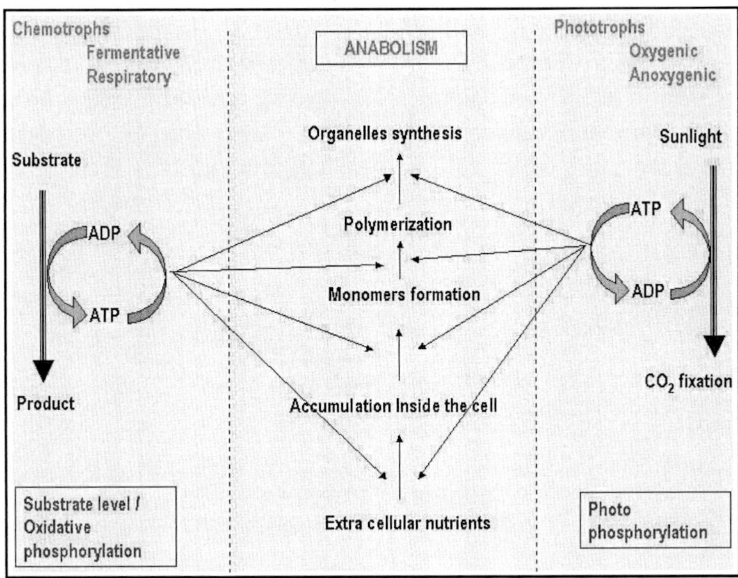


*Energy generation by heterotrophs*

Heterotrophs use variety of carbon sources. Glucose is being the simple and wide variety of microbes prefer it. The glucose can be taken up by bacterium through diffusion and can be readily utilized. There are three possible pathways available in bacteria to use glucose.

A. Embden-Meyerhof pathway
B. Phosphoketolase pathway
C. Entner - Doudoroff pathway.

All these path ways are fermentative type and substrate level phosphorylation occurs.

## Substrate Level Phosphorylation (Fermentation)

The EMP pathway, phosphoketolase pathway and ED pathway end with one or two ATP synthesis by substrate level phosphorylation. There won't be any external source of electron acceptor will come in these reactions.

*A. Embden-Meyerhof pathway*

This is the pathway of glycolysis most familiar and common to most of the organisms. The pathway is operated by yeast to produce alcohol and lactic acid bacteria to produce lactic acid and several organic acids, gases, fatty acids, and alcohols.

The pathway is as follows :

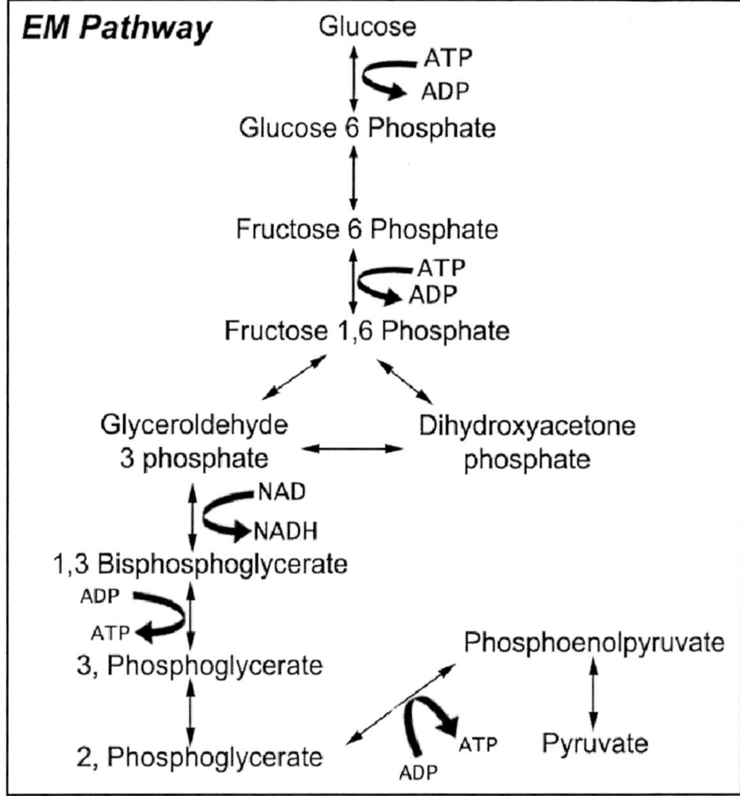


The overall reaction is

Glucose → 2 pyruvate + 2 ATP + 2 $NADH_2$

After pyruvate is formed, if the organism is a *respirative* type, the pyruvate will go to *Krebs cycle* and if the organism is *fermentative,* the *reduction* process ends with organic acids, alcohols etc.

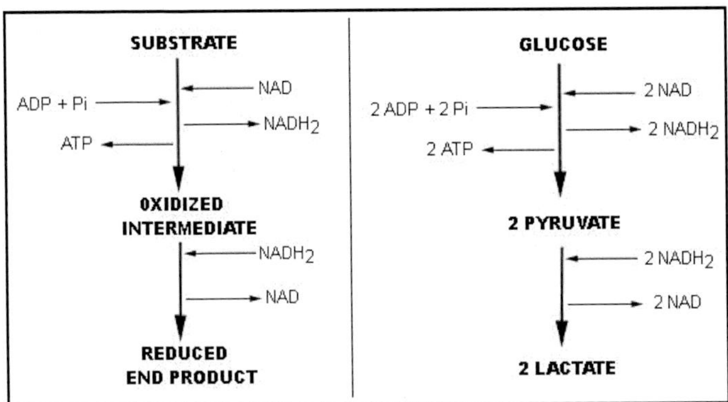


*A model fermentation:* After an intermediate product, a reduction takes place in fermentation, whereas, if respiration, $CO_2$ will be formed by complete oxidation through Krebs's cycle

*Note :* After pyruvate, the reduction process leads to *fermentation* and complete oxidation leads to *respiration*

The Embden Meyorhof pathway can lead to a wide array of end products depending on the pathways taken in the reductive steps after the pyruvate formation.

The following are some of the such fermentations :

| Fermentation | End products | Model organism |
|---|---|---|
| Homolactic fermentation | Lactic acid | *Lactobacillus* |
| Mixed acid fermentation | Lactate, acetate, formate, succinate | *Enterobater* |
| Butyric acid fermentation | Butric acid, acetone | *Clostridium acetobutylicum* |
| Propionic acid fermentation | Propionic acid | *Propionibacterium* |
| Alcohol fermentation | Ethanol | *Saccharomyces* |

*B. Phosphoketolase pathway (Heterolactic pathway)*

The phosphoketolase pathway is distinguished by the key cleavage enzyme **phosphoketolase,** which cleaves pentose to glyceroldehyde 3 phosphate and acetyl phosphate. The path way ends with ethanol and lactic acid. Ex. *Lactobacillus, Leuconostoc.* The overall reaction is

Glucose → 1 lactate + 1 ethanol + 1 $CO_2$ + 1 ATP

This pathway is useful in the dairy industry for preparation of kefir (fermented milk), yogurt, etc.

The pathway is as follows :

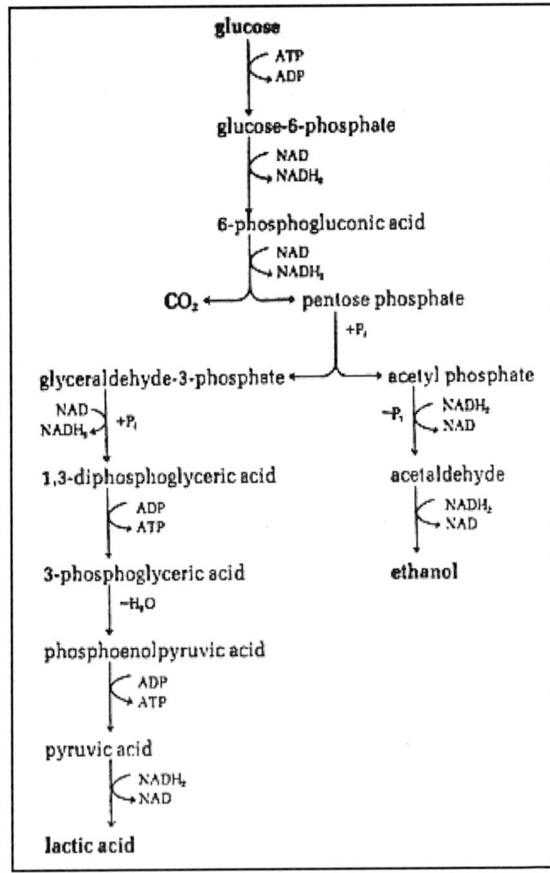

*C. Entner – Doudoroff pathway*

Only few bacteria like, *Zymomonas mobilis* employ the ED pathway. The path way is as follows:

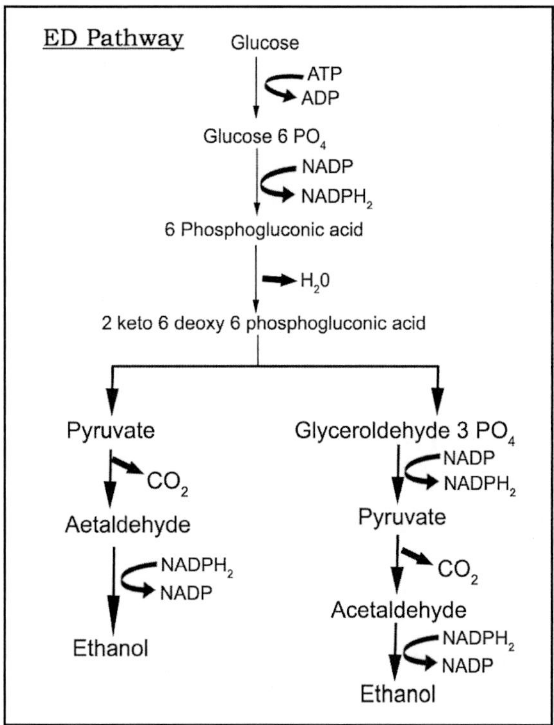


The overall reaction is

Glucose → 2 ethanol + 2 $CO_2$ + 1 ATP

The alcohol productivity of *Zymomonas* is higher than yeast because of this fermentative pathway.

*Note :* All the three pathways are end with 1 or 2 ATP by substrate level phosphorylation by means fermentation

## Oxidative Phosphorylation (Respiration)

If the organism is a respiratory type (that means complete oxidation of glucose), it needs following essential metabolic components for their respiration and oxidative phosphorylation.

a. *Tricarboxylic acid cycle (also known as citric acid cycle or Kreb's cycle)* : The pyruvate formed during glycolysis will be completely oxidized to 3 $CO_2$ by the use of this cycle. The metabolic pathway is as follows :

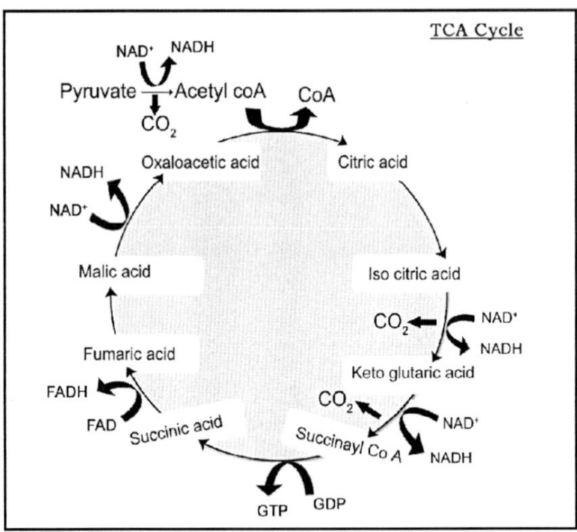


During oxidation of one pyruvate through TCA cycle, 4 $NADH_2$, 1 $FADH_2$ and 1 GTP are produced along with 3 $CO_2$.

b. *A membrane associated Electron Transport System (ETC)* : The electron transport chain is a sequence of electron carriers transport the electrons to a terminal electron acceptor. During this flow of electron in the membrane, a proton motive force across the membrane leads to formation ATP (is referred as electron transport phosphorylation).

c. *An outside electron carrier:* for aerobic respiration, $O_2$ is the terminal electron acceptor and reduced to $H_2O$. This is normal for higher organisms. But in anaerobic bacteria, the terminal electron acceptor may be of nitrite, nitrate, sulphate or carbon dioxide.

The following diagram shows the flow of electron transport in membrane and ATP fomation due to proton motive force.

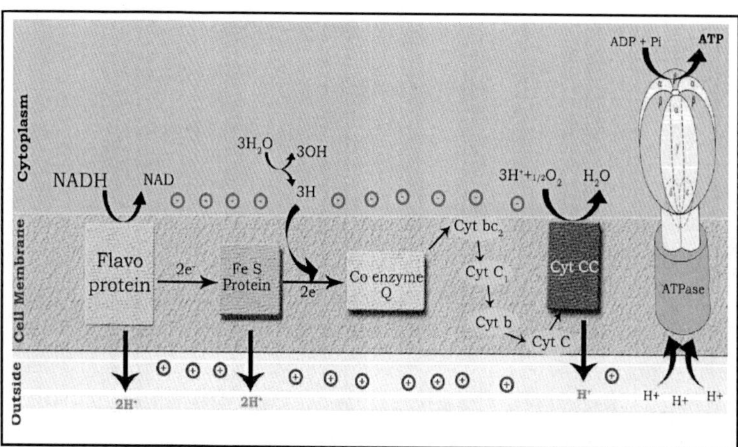


The table shows some aerobic and anaerobic respirations with specific examples:

| Terminal electron acceptor | End product | Process name | Organism |
|---|---|---|---|
| $O_2$ | $H_2O$ | Aerobic respiration | *Streptomyces* |
| $NO_3$ | $NO_2$, $N_2$ | Denitrification | *Pseudomonas denitrificans* |
| $SO_4$ | S or $H_2S$ | Sulphate reduction | *Desulfovibrio desulfuricans* |
| Fumarate | Succinate | Anaerobic respiration | *Escherichia* |
| $CO_2$ | Methane ($CH_4$) | Methanogenesis | *Methanococcus* |

In aerobic organisms, the terminal electron acceptor will be of $O_2$. In some anaerobic organisms, after the electron transport chain, instead of $O_2$, some inorganic compounds like sulphate, nitrate or some organic compounds like fumarate act as terminal electron acceptor. Such type of respiration is referred as *anaerobic respiration* and the normal $O_2$ mediated respiration is referred as *aerobic respiration*.

The above table shows some anaerobic respiration with some terminal electron acceptors. The process is named based on the compounds as *sulphur reduction, denitrification* and *methanogenesis*.

## Energy generation by autotrophs

Autotrophs use $CO_2$ as their sole carbon source. There are two types such as *photoautotrophs* and *chemoautotrophs*.

Photoautotrophs use light as energy source and $CO_2$ as carbon source.

Chemoautotrops use chemicals (specially inorganic) as energy source and $CO_2$ as carbon source.

## Energy and carbon assimilation by photoautotrophs

Phototrophs use sunlight to produce ATP through phosphorylation, referred as *photophosphorylation*. The phototrophs convert the light energy to chemical energy (ATP) through the process called *photosynthesis*.

Photosynthesis is a type of metabolism in which catabolism and anabolism occur as sequence. The catabolic reaction (energy generating process) of photosynthesis is *light reaction* in which the light energy is converted to chemical energy (ATP) and electrons or reducing powers (NADPH). The anabolic reaction (macromolecule synthesis) of photosynthesis is *dark reaction* in which $CO_2$ is converted to organic molecules (carbohydrates), which is also called as *$CO_2$ fixation*.

For conversion of light energy to ATP, the bacteria possess *light harvesting pigments*. They are *chlorophyll a, carotenoids, phycobiliproteins* (which are present in cyanobacteria) and *bacteriochlorophyll* (which are present in purple sulphur bacteria).

In bacteria, there are two types of light reactions (conversion of light to ATP) and two types of $CO_2$ fixation occur.

## Photophosphorylation

For photophosphorylation, *a light harvesting pigments, a membrane electron transport chain, source of electron (electron donor)* and *ATPase* enzymes are required.

Two types of photophosphorylations occur during photosynthesis. *Cyclic photophosphorylation and non-cyclic photophosphorylation.*

- In plant and cyanobacteria, both cyclic and non-cyclic photophosphorylation occur whereas in purple bacteria, the cyclic photophosphorylation only occurs.
- In plant and cyanobacteria, the electron source is water, by *photolysis*, $H_2O$ split into $H^+$ and $O_2$ and during the process, $O_2$ is evolved and referred as *oxygenic photosynthesis*
- Since, the sulphur bacteria is an anaerobic bacterium, they use $H_2S$ instead of $H_2O$ as electron donor. Since, there won't be any $O_2$ evolution during photosynthesis, referred as *anoxygenic photosynthesis*.

Difference between plant and bacterial photosynthesis

| Organisms | Plant photosynthesis | Bacterial photosynthesis |
|---|---|---|
| | plants, algae, cyanobacteria | purple and green bacteria |
| Type of chlorophyll | chlorophyll a absorbs 650-750nm | bacteriochlorophyll absorbs 800-1000nm |
| Photosystem I (*cyclic photophosphorylation*) | present | present |
| Photosystem II (*noncyclic photophosphorylation*) | present | absent |
| Produces $O_2$ | Yes (*Oxygenic*) | No (*Anoxygenic*) |
| Photosynthetic electron donor | $H_2O$ | $H_2S$, other sulfur compounds or certain organic compounds |

The oxygenic photophosphorylation is as follows :

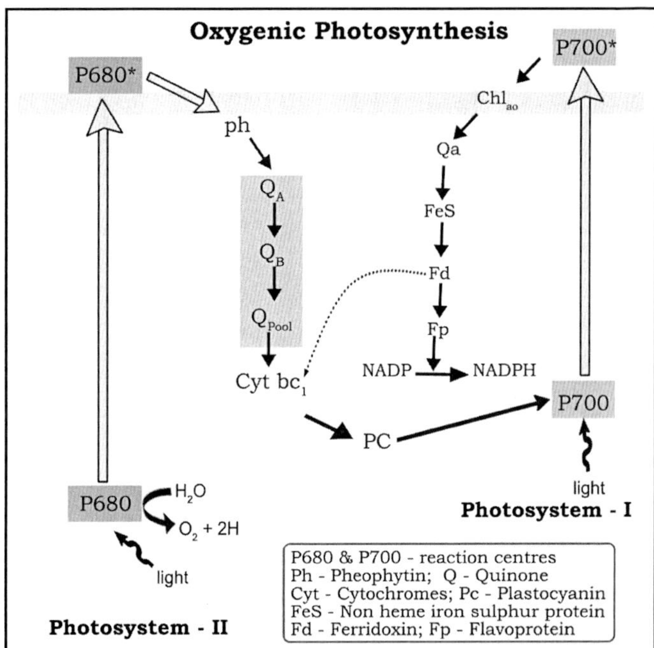


The end product of the light reaction is ATP, NADPH and $O_2$. The ATP and NADPH, the energy and electron sources thus produced were used for *dark reaction.*

The anoxygenic photo phosphorylation will take palce as in the image and the The end product of the light reaction is ATP, NADPH and Sulphur. The ATP and NADPH, the energy and electron sources thus produced were used for *dark reaction.*

## Dark reaction ($CO_2$ fixation)

The dark reaction in which the ATP and NADPH were used as energy and electron sources to fix the $CO_2$ as carbohydrates. The pathway involved in the dark reaction is *calvin cycle,* by which the $CO_2$ is fixed as *phosphoglyceic acid* and lead to formation of many sugars. The enzyme RuBiSCO is the key enzyme for this process.

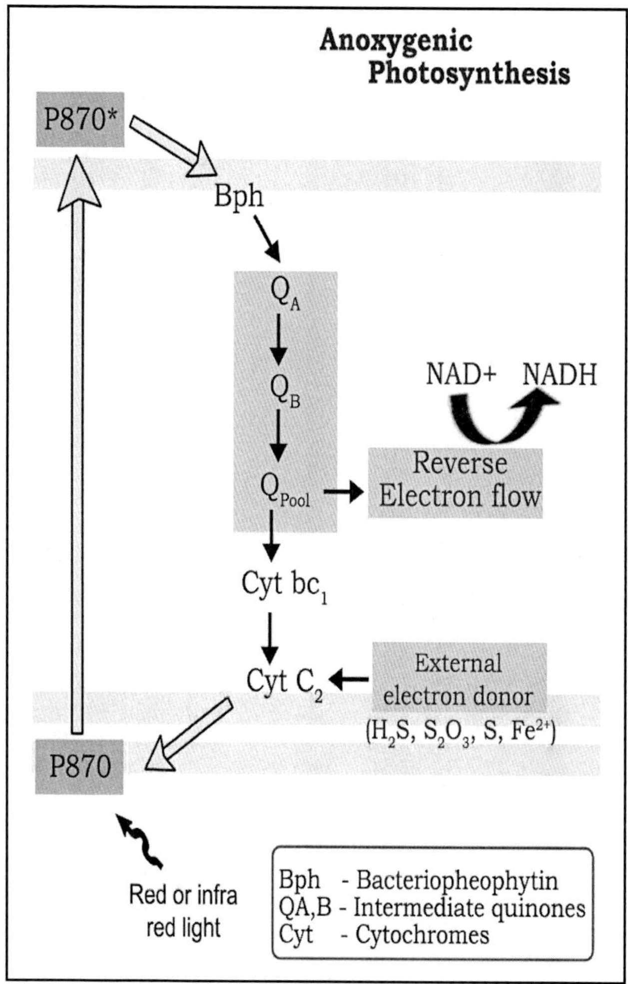


The following pathway shows the Calvin cycle and the formation of key monomers for anabolic reactions such as hexose phosphate - *polysaccharides*; pyruvic acid - *amino acid* and *fatty acid*; pentose phosphate - *DNA* and *RNA*.

## Another way of $CO_2$ fixation by phototrophs

In phototrophs, the electron and energy were derived form sunlight and carbon from $CO_2$ fixation through Calvin

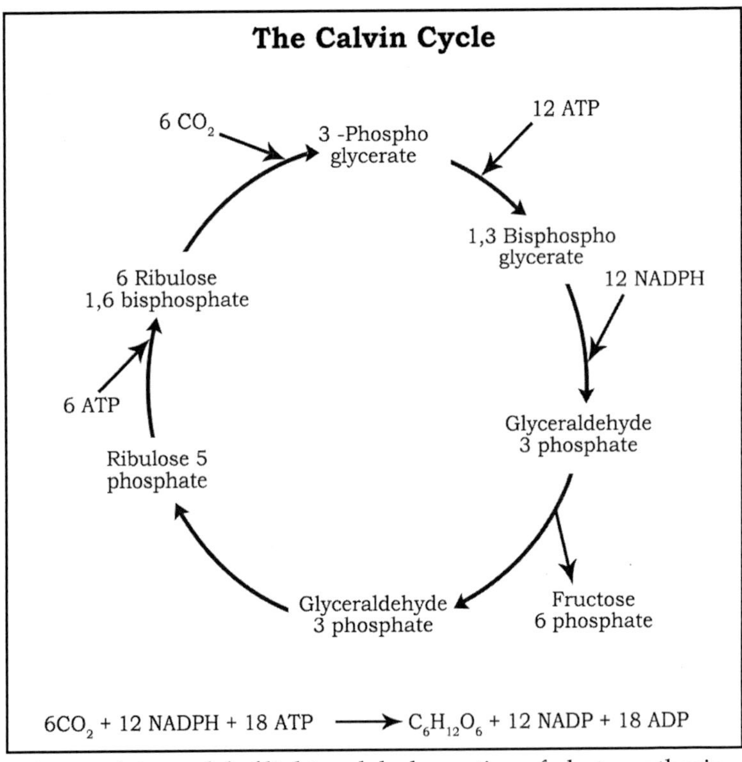


A complete model of light and dark reaction of photosynthesis

cycle. But some bacteria may derive electron and energy from sunlight and fixes $CO_2$ by some other path way not, Calvin cycle. The example is Photosynthetic green bacteria (*Chlorobium*). They derive NADPH and ATP through cyclic phosphorylation, but $CO_2$ fixation is by *reverse TCA cycle*. Since TCA cycle is *amphibolic pathway* (referring the cycle can operate in both the directions), it can also be used to fix the carbon-di-oxide if operated reversely. The pathway is as follows:

Another way of $CO_2$ fixation is by methanogens. They use $CO_2$ as terminal electron acceptor and forms $CH_4$ (methane). They also fix by acetyl CoA pathway for fixing $CO_2$.

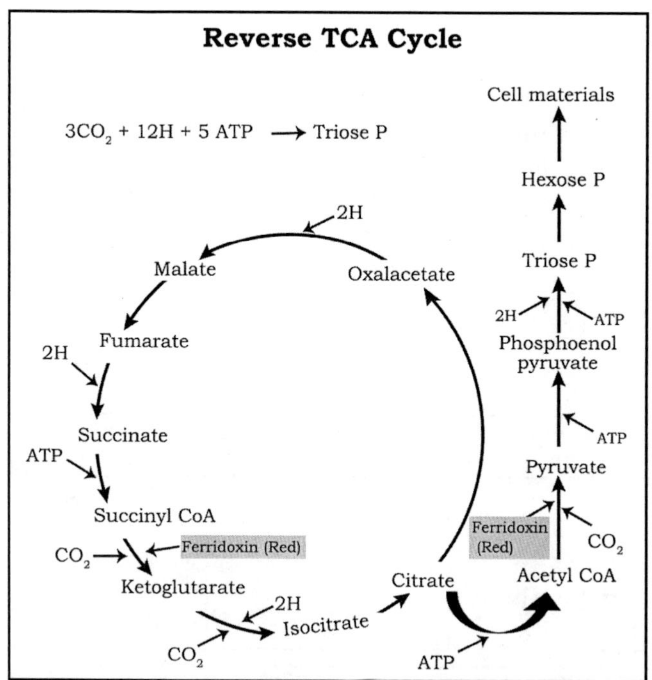


## Synopsis

| Organism | Light reaction/ ATP generation | Dark reaction/ $CO_2$ fixation |
|---|---|---|
| Cyanobacteria (*Nostoc*), plant and alga | • Cyclic and non-cyclic photophosphorylation<br>• Oxygenic photosynthesis | Kalvin cycle |
| Purple bacteria (*Chromatium*) | • Cyclic and non-cyclic photophosphorylation<br>• Oxygenic photosynthesis | Kalvin cycle |
| Green bacteria (*Chlorobium*) | • Cyclic and non-cyclic photophosphorylation<br>• Oxygenic photosynthesis | Reverse TCA cycle |

## Energy and carbon assimilation by Chemoautotrophs

Since the chemoautotrophs use inorganic chemicals for their energy and electron source, they are referred as *chemolithotrophs* or *chemolithotrophic autotrophs*.

These organisms remove electron from an inorganic substance and put them through electron transport chain for ATP synthesis (through electron transport phosphorylation). At the same time, the electrons also flow through *reverse electron transport chain* and with the end product of NADPH. These ATP and NADPH are used for $CO_2$ fixation through Calvin cycle.

These bacteria are obligate aerobic organisms. Some examples of the chemolithotrophs are as follows:

Groups of chemolithotrophs

| Physiological group | Energy source | Oxidized end product | Organism |
|---|---|---|---|
| Hydrogen bacteria | $H_2$ | $H_2O$ | *Alcaligenes, Pseudomonas* |
| Nitrifying bacteria | $NH_3$ | $NO_2$ | *Nitrosomonas* |
| Nitrifying bacteria | $NO_2$ | $NO_3$ | *Nitrobacter* |
| Sulfur oxidizing bacteria | $H_2S$ or S | $SO_4$ | *Thiobacillus, Sulfolobus* |
| Iron oxidizing bacteria | $Fe^{2+}$ | $Fe^{3+}$ | *Gallionella, Thiobacillus* |

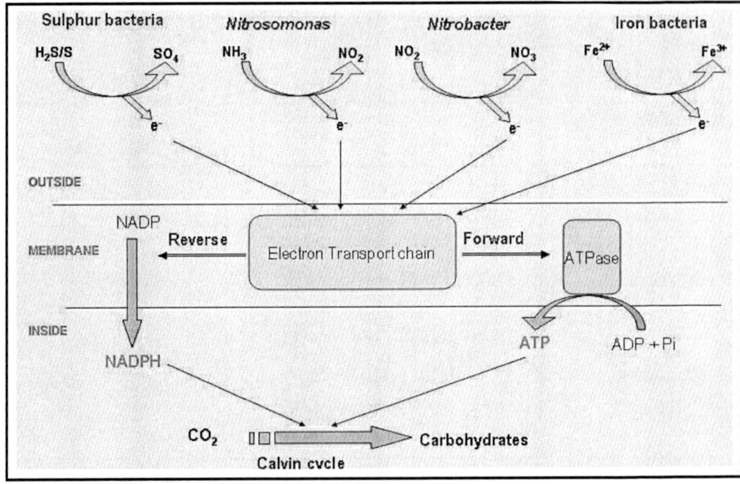


The above diagram sows energy generation and $CO_2$ fixation by different chemolithotrophs.

CHAPTER

11

# Control of Microorganisms

Sterilization refers the complete removal or elimination of microorganisms from an environment. Physical and chemical agents are available to remove the microorganisms from the environment. Heat, irradiation and filtration are the three physical agents and various chemicals are available for sterilization.

## Terms related to control of microorganisms

*Cide or cidal* means to kill the microorganism (Ex. Bactericidal refers the killing of bacteria).

*Static or stasis* means inactivation or inhibition, but not killing. They prevent multiplication of the organism. (Ex. Fungistatic refers the inhibition of growth of fungi.).

*Sepsis* refers break down of a living tissue by organisms and is accompanied by inflammation and pus formation.

*Antispetic* is an agent applied externally on living tissues to kill or inhibit the growth of the organisms.

*Disinfectant* is an agent applied externally on inanimate objects (not living tissues) to destroy harmful pathogens in their vegetative stage.

*(Antiseptics are milder than disinfectants because to avoid the side effects)*

*Sanitation* refers removal of organisms from a location by cleaning but not by sterilization.

## A. Physical Agents

### I. Heat

*Principle :* At high temperature, all the macro molecules lose their structure and ability to function called as denaturation. Similarly, the heat also coagulate the protein and oxidize the elements lead for lethality.

The heat susceptibility of the organism should know before the heat treatment to be applied. For this, the thermal death time and thermal death points are essential.

*Thermal death time* refers the time to kill a suspension of organisms at a specific temperature.

*Thermal death point* refers the lowest temperature to kill a suspension in a given time.

*Decimal reduction time* (D time) – Time to kill 90% or 1 log unit of a population at a given temperature

Based on the above parameters, the time and temperature to kill the microorganism from an environment is derived.

4 or 5 different ways of sterilization are available to kill the microorganisms using heat.

1. *Direct flaming :* showing the objects or equipments directly to the flame leads the sterilization. This is also called as incineration. The inoculation needles, spread rods and forceps can be sterilized using this technique.
2. *Boiling :* In this method, only the vegetative forms are killed in few minutes whereas the spores are not.
3. *Dry heat :* Causes oxidation of cells. By using electrical coils, heat is generated and used for sterilization. Normally high temperature and long time are required

for complete sterilization using this method. 160-180°C for 2- 3 hours are required for complete sterilization. Hot air oven is an instrument used for dry heat sterilization. It is a double walled (inner and outer walls) instrument protected with asbestos sheets to avoid heat loss. Thermostat is present to control the temperature and timer is also present to turn off automatically after particular time. The glasswares like conical flasks, beakers, petri dishes, pipettes can be sterilized using this technique.

4. *Moist heat :* causes denaturation and coagulation of protein. To kill vegetative structure of bacteria, yeast and molds - 80°C for 5-10 min; to kill mold spores - 80°C for 30 min; bacterial spores - 121°C, 15 lbs/ sq.inch pressure for 15 min. are the optimum moist heat. The autoclave is used for sterilization using moist heat under pressure. Autoclave is made up of double walled steel plates with air tight lids. Electrical coil submerged in water is provided to generate the stream. To measure the pressure, a pressure gauge and for safety, a safety valve is also attached in the lid. If the steam pressure inside the closed vessel is increased to 15 lb/sq. inch, the temperature will raise to 121.6°C. Keeping this condition for 15 – 20 min. will kill all the vegetative and spore forms of microbes.

(The moist heat has more penetration power than dry heat and leads faster reduction in number of living organism and under pressure, the penetration of heat will be more in the autoclave).

Tyndallization

Refers intermittent or fractional sterilization. The subsequent cooling and heating by steam for 3 days will remove the germs and their spores refers the tyndallization. Ex. Soil which contains diversified microbes with spores. The tyndallization will remove all the vegetative cells and also the spores. The normal sterilization by autoclave will eliminate the vegetative cells only and not spores.

Pasteurization

Refers removal of undesired microorganisms without affecting the beneficial microorganisms, oduour and taste. The pasteurization of milk is done to eliminate the pathogenic microorganisms which cause tuberculosis, brucellosis, Q fever, typhoid etc. In broad, the pasteurization is done to kill *Salmonella* and *E. coli* like organisms.

*Flash pasteurization :* holding the substance at 71°C for 15 seconds and rapid cooling referred as flash pasteurization.

*Bulk pasteurization :* holding the substance at 63-65°C for 30 min and slow cooling referred as bulk pasteurization

*Tyndallization > sterilization (by autoclave) > pasteurization* is the order of degree of strength in terms of removal of microorganisms.

*II. Radiation*

Microwaves, ultra violet rays, gamma rays, electrons have the power to kill the microorganisms. Among them, UV light (non-ionizing radiation) and gamma rays (ionizing radiation) are commonly used for sterilization.

UV radiation

The wave length considered to be between 220 - 300 nm is used to control the microorganisms. It has sufficient energy to form *dimers* (bridges) between two adjacent pyrimidines in the DNA and thereby affect the DNA replication leads to death. This light is highly useful to disinfect surfaces, air, and water, which do not absorb the light. Laminar air flow chamber - A clean bench used to handle the microbes in a microbe free environment. It has two methods of sterilization. Radiation by UV and filtration by HEPA filters (High efficiency particulate filters), a microbe free air will be present in side.

Ionizing radiation

Ionizing radiation is electromagnetic radiation of sufficient energy to produce ions like electron, hydroxyl ions, protons, etc. which can alter the biopolymers such as DNA and protein causing break of polymers. $\gamma$-rays produced from $^{60}Co$ and $^{137}Cs$ are useful for sterilization of the plastic wares, gloves, heat labile plastic containers, poly bags, tissue grafts, plastic syringes etc. In few cases, food were also sterilized by $\gamma$-rays.

*III. Filtration*

Heat is the most common and effective way of sterilization, some heat sensitive liquids cannot be sterilized by heat. An alternate technique for the sterilization is by filters. A filter is too small for a passage of microorganisms but large enough to passage of liquids.

There are three types of filters

a. *Pyrex glass or sintered glass filters :* Filter glass made with powdered glass disc pore size of 0.5 to 2 mm discs. These discs were fit into funnels and used for filtration. These were used by suction pumps for faster flow rate.

b. *Seitz filters :* The filter mat is made up of asbestos – cellulose mixture.

c. *Membrane filters :* They are composed of polymers with high tensile strength such as cellulose acetate, cellulose nitrate or polysulfonate, manufactured in such a way that they contain a large number of tiny holes. By adjusting the polymerization conditions during manufacture, the size of holes in the membrane can be precisely controlled. They trap the microbes on the surface and allow the microbe free solution to pass.

*B. Chemical agents*

Different chemicals like halogens, heavy metals, phenols, alcohols, ethanol are used for sterilization.

The following are different classes of chemicals involved in sterilization.

1. *Phenol and Phenolics :* 5% aqueous solutions kill vegetative forms. Some of their derivatives are Lysol, Dettol, Cresol etc.. They alter the selective permeability of the cytoplasmic membrane of organisms.

2. *Alcohols :* They are bactericidal and fungicidal but not sporicidal. They denature proteins and are solvents of lipids. 60 - 90% is the range for this activity. Water is required for the lethal effect.

3. *Halogens :* Cause oxidation and direct halogenation of proteins thus inhibiting the activity of proteins e.g. hypochlorite, iodophores etc.

4. *Surfactants :* Surface active agents, good wetting and solubilizing agents e.g. Soap, Detergent etc. They have both hydrophilic and hydrophobic groups (ambivalent) to dissolve compounds.

5. *Alkylating agents :* They substitute alkyl groups for hydrogen of reactive groups in nucleic acids and proteins, causing disruption of metabolic pathways e.g. Formaldehyde, glutaraldehyde, ethylene oxide (gaseous agent).

6. *Heavy metals :* reacts with sulfhydryl group (SH) of enzyme and make them inactive. Mercury, copper and silver are commonly used heavy metals.

7. *Antibiotics* : Antibiotics are low molecular weight molecules produced as secondary metabolites by microorganisms, which kill or inhibit the other microorganisms. The following table shows the summarized examples of common antibiotics and their action.

| Chemical | Use | Mode of action |
|---|---|---|
| **Antisepsis** | | |
| Alcohol | Skin | Lipid solvent and protein denaturant |
| Phenol compounds | Soaps, lotions, cosmetics, body deodorants | Disrupt cell membrane |
| Cationic detergents (Ex. Benzalkonium chloride) | Soap, lotion | Phospholipid interaction |
| 3% hydrogen peroxide | Skin | Oxidizing agent |
| Iodine containing iodophors | Skin | Iodinates tyrosine residues; oxidizing agent |
| Silver nitrate | Eyes of new born to prevent blindness from *Nesseria gonorrhoea* | Protein precipitant |
| **Disinfectants** | | |
| Alcohol | Medical instruments, food, dairy instruments, lab surface | Lipid solvent & protein denaturant |
| Cationic detergents | Medical instruments, food, dairy instruments, lab surface | Interact with phospholipid |
| Chlorine gas | Disinfectant for purification of water supplies | Oxidizing agent |
| Chlorine compounds (Chloramines, sodium hypochlorite, chlorine dioxide) | Disinfectant for food and dairy industry, water supplies | Oxidizing agent |

*Contd...*

*Table Contd...*

| | | |
|---|---|---|
| Copper sulphate | Algicide in swimming pools | Protein precipitant |
| Ethylene oxide (gas) | Sterilant for temperature sensitive lab material like plastics | Alkylating agent |
| Formaldehyde | 3-8% used as surface disinfectant; 37% (referred as formalin) as preservatives | Alkylating agent |
| Gluteraldehyde | 2% solution | Alkylating agent |
| 3% hydrogen peroxide | Medical instruments | Oxidizing agent |
| Mercuric chloride | For laboratory surface, plant samples | Protein denaturation |
| Ozone | Drinking water | Oxidizing agent |
| **Preservatives** | | |
| Salt and sugar solutions | Preservative | Osmotic pressure |

| S. No. | Antibiotics | Source | Effective against | Mode of action |
|---|---|---|---|---|
| 1. | Penicillin | *Pennicillium notatum, P. crysogenum* | Gram +ve bacteria | Inhibit the cell wall (Peptidoglycon layer) synthesis |
| 2. | Amphicillin | | Gram +ve bacteria | Inhibit the cell wall (Peptidoglycon layer) synthesis |
| 3. | Streptomycin | *Streptomyces griseus* | G + and G-ve organisms | Inhibit the translation process of protein synthesis |
| 4. | Erythromycin | *Streptomyces erythreus* | G + and G-ve | Inhibit the translation process of protein synthesis |

*Contd....*

| 5. | Polymixin | *Bacillus polymyxa* | Gram - ve | Damages Cytoplasm |
|---|---|---|---|---|
| 6. | Tetracycline | *Streptomyces* sp. | G+ and G- ve | Inhibit the translation process of protein synthesis |
| 7. | Chloremphe necol | *Streptomyces* sp. | Bacteriostatic | Inhibit the Protein synthesis |

CHAPTER 12

# Microbial Genetics — Basic Concepts

*Genetics* is the study of what genes are, how they carry information, how their information is expressed, and how they are replicated and passed to subsequent generations or other organisms.

DNA in cells exists as a double-stranded helix; the two strands are held together by hydrogen bonds between specific nitrogenous base pairs: A-T and C-G.

A gene is a segment of DNA, a sequence of nucleotides, that code for a functional product, usually a protein.

When a gene is expressed, DNA is transcribed to produce RNA; mRNA is then translated into proteins. This is also referred as Central Dogma of Life. The DNA in a cell is replicated before the cell divides, so each daughter cell receives the same genetic information.

## Genotype and Phenotype

Genotype is the genetic composition of an organism—its entire DNA. Phenotype is the expression of the genes—the proteins of the cell and the properties they confer on the organism.

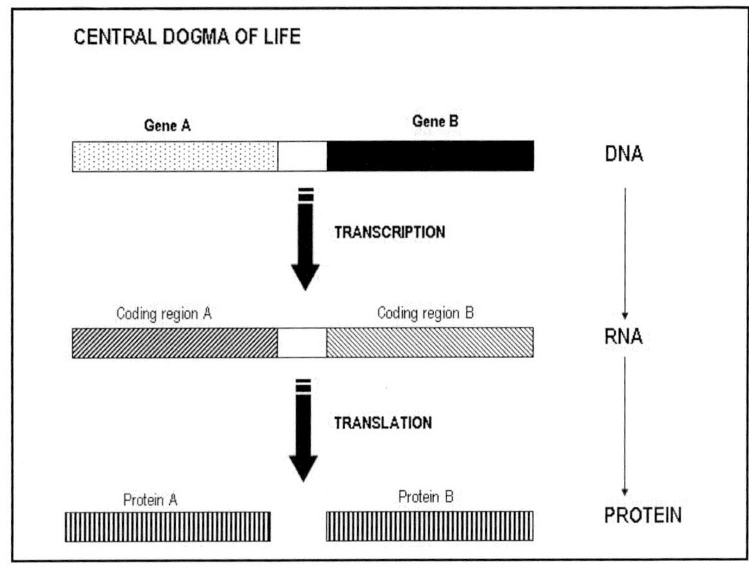


## DNA and Chromosomes

The DNA in a chromosome exists as one long double helix associated with various proteins that regulate genetic activity.

Bacterial DNA is circular; the chromosome of *E. coli,* for example, contains about 4 million base pairs and is approximately 1000 times longer than the cell.

Genomics is the molecular characterization of genomes. Information contained in the DNA is transcribed into RNA and translated into proteins.

## Enzymes that associated with DNA

1. *DNA polymerase :* An enzyme that synthesize a new strand of DNA in $5' \rightarrow 3'$ direction using anti parallel strand as template.

2. *Restriction Endonuclease :* An enzyme which recognizes and makes double strand DNA breaks at specific sequence of DNA.

3. *RNA polymerase :* An enzyme that synthesize RNA in $5' \rightarrow 3'$ direction using anti parallel DNA strand as template.
4. *DNA ligase :* An enzyme that lygate or joint double strand DNA fragments.
5. *DNA gyrase :* This enzyme introduce super coiling of DNA in prokaryotes (also referred as Topoisomerase II).
6. *Topoisomerase I :* An enzyme which removes the super coiling of DNA (Super coiling refers the highly twisted form of DNA).

## DNA Replication

During DNA replication, the two strands of the double helix separate at the replication fork, and each strand is used as a template by DNA polymerases to synthesize two new strands of DNA according to the rules of nitrogenous base pairing.

The result of DNA replication is two new strands of DNA, each having a base sequence complementary to one of the original strands. Because each double-stranded DNA molecule contains one original and one new strand, the replication process is called as semiconservative. DNA is synthesized in one chemical direction called $5' \rightarrow 3'$ ($5'$ is phosphate end; $3'$ is hydroxyl end of deoxyribose). At the replication fork, the leading strand is synthesized continuously and the lagging strand, discontinuously. DNA polymerase proof reads new molecules of DNA and removes mismatched bases before continuing DNA synthesis. Errors only occur ~1 time for every $10^{10}$ bases added. Each daughter bacterium receives a chromosome identical to the parent's.

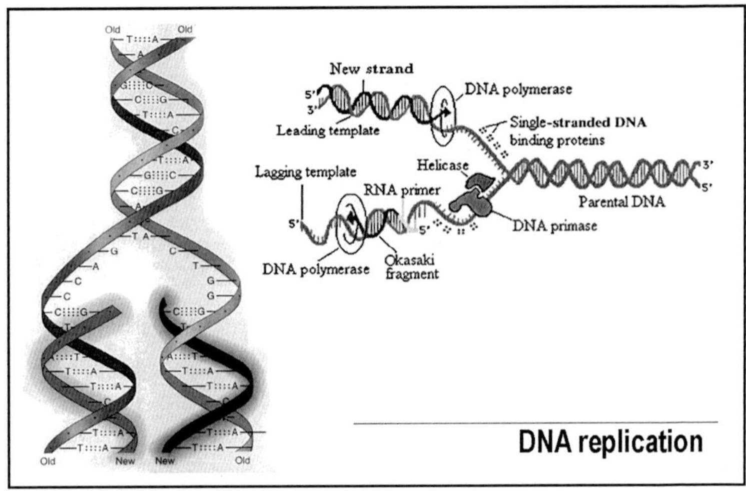


DNA replication

## Protein Synthesis

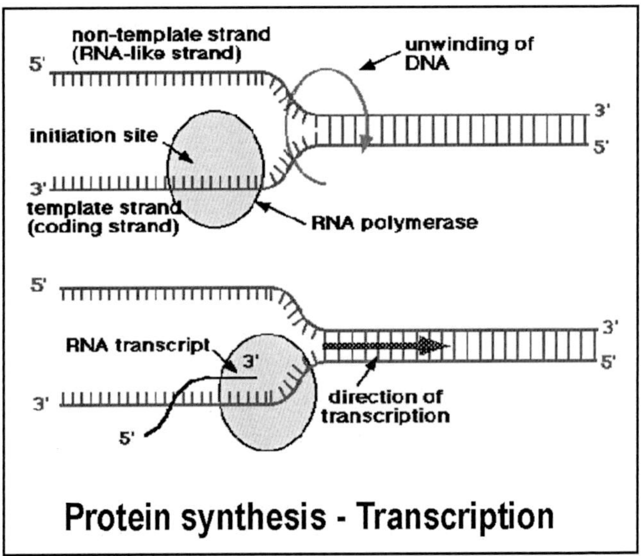


Protein synthesis - Transcription

## Transcription

During transcription, the enzyme RNA polymerase synthesizes a strand of RNA from one strand of double-stranded DNA, which serves as a template.

RNA is synthesized from nucleotides containing the bases A, C, G, and U, which pair with the bases of the DNA sense strand. The starting point for transcription, where RNA polymerase binds to DNA, is the promoter site; the region of DNA that is the endpoint of transcription is the terminator site; RNA is synthesized in the 5' —> 3' direction.

## Translation

Translation is the process in which the information in the nucleotide base sequence of mRNA is used to dictate the amino acid sequence of a protein.The mRNA associates with ribosomes, which consist of rRNA and protein.

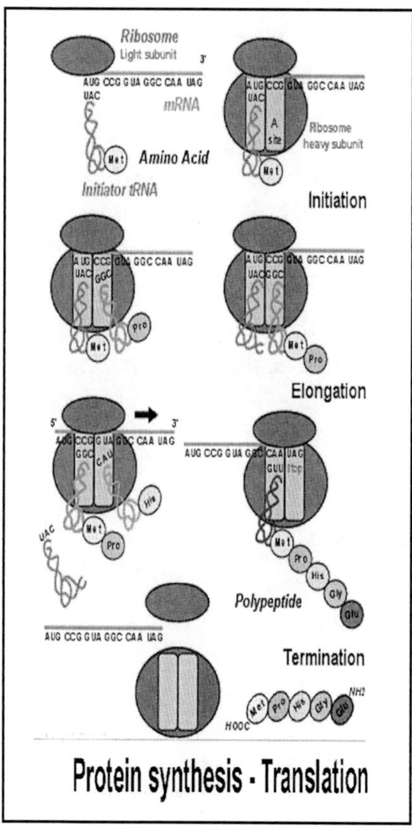


Protein synthesis - Translation

Three-base segments of mRNA that specify amino acids are called *codons.* The *genetic code* refers to the relationship among the nucleotide base sequence of DNA, the corresponding codons of mRNA, and the amino acids for which the codons code. The genetic code is degenerate; that is, most amino acids are coded for by more than one codon. Of the 64 codons, 61 are sense codons (which code for amino acids), and 3 are nonsense codons (stop codons) which do not code for amino acids and are stop signals for translation. The start codon, AUG, normally codes for methionine (formylmethionine at the beginning of a protein). Specific amino acids are attached to molecules of tRNA. Another portion of the tRNA has a base triplet called an anticodon. The base pairing of codon and anticodon at the ribosome results in specific amino acids being brought to the site of protein synthesis. The ribosome moves along the mRNA strand as amino acids are joined to form a growing polypeptide; mRNA is read in the 5' —> 3' direction. Translation ends when the ribosome reaches a stop codon on the mRNA.

## Bacterial Genome Organization

DNA molecules that replicate as discrete genetic units in bacteria are called replicons. In some *Escherichia coli* strains, the chromosome is the only replicon present in the cell. Other bacterial strains have additional replicons, such as plasmids and bacteriophages.

## Chromosomal DNA

Bacterial genomes vary in size from about $0.4 \times 10^9$ to $8.6 \times 10^9$ daltons (Da), some of the smallest being obligate parasites (*Mycoplasma*) and the largest belonging to bacteria capable of complex differentiation such as *Myxococcus*. The amount of DNA in the genome determines the maximum amount of information that it can encode. Most bacteria have a haploid genome, a single chromosome consisting of a circular, double stranded DNA molecule. However linear chromosomes have been found in Gram-positive *Borrelia* and *Streptomyces* spp., and one linear and one circular chromosome is present in the Gram-negative bacterium *Agrobacterium tumefaciens.*

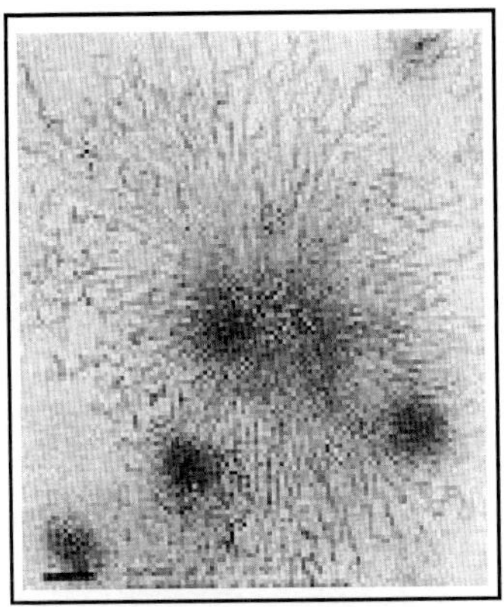

The single chromosome of the common intestinal bacterium *E. coli* is 3 x $10^9$ Da (4,500 kilobase pairs [kbp]) in size, accounting for about 2 to 3 percent of the dry weight of the cell. The *E. coli* genome is only about 0.1% as large as the human genome, but it is sufficient to code for several thousand polypeptides of average size (40 kDa or 360 amino acids).

The chromosome of *E. coli* has a contour length of approximately 1.35 mm, several hundred times longer than the bacterial cell, but the DNA is supercoiled and tightly packaged in the bacterial nucleoid. The time required for replication of the entire chromosome is about 40 minutes, which is approximately twice the shortest division time for this bacterium. DNA replication must be initiated as often as the cells divide, so in rapidly growing bacteria a new round of chromosomal replication begins before an earlier round is completed. At rapid growth rates there may be four chromosomes replicating to form eight at the time of cell division, which is coupled with completion of a round of chromosomal replication. Thus, the chromosome in rapidly growing bacteria is replicating at more than one point. The replication of chromosomal DNA in bacteria is complex and involves many different proteins.

## Plasmids

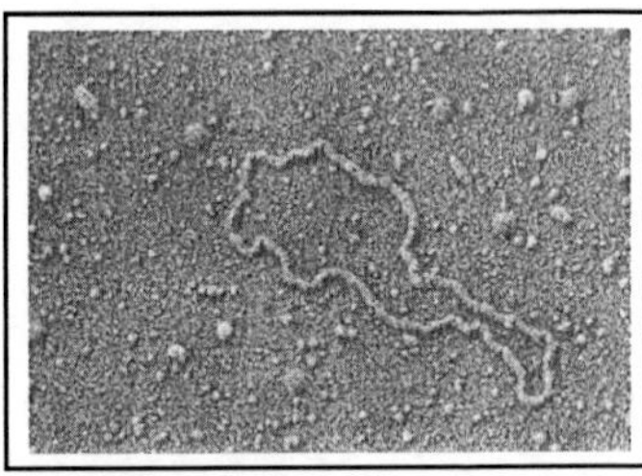

Plasmids are replicons that are maintained as discrete, extra chromosomal genetic elements in bacteria. They are usually much smaller than the bacterial chromosome, varying from less than 5 to more than several hundred kbp, though plasmids as large as 2 Mbp occur in some bacteria. Plasmids usually encode traits that are not essential for bacterial viability, and replicate independently of the chromosome. Most plasmids are supercoiled, circular, double-stranded DNA molecules, but linear plasmids have also been demonstrated in *Borrelia* and *Streptomyces*.

Many plasmids control medically important properties of pathogenic bacteria, including resistance to one or several antibiotics, production of toxins, and synthesis of cell surface structures required for adherence or colonization. Plasmids that determine resistance to antibiotics are often called R plasmids (or R factors) and the plasmids responsible for conjugation are F plasmids.

## Gene regulation & expression

The genes are always present in the bacterial chromosomes but their expression is always controlled. Here the gene expression refers the mRNA and protein synthesis. The expression of the gene was systematically controlled by some proteins. Whenever needed, the genes will be expressed to get the protein (i.e., enzymes).

Regulating protein synthesis at the gene level is energy-efficient because proteins are synthesized only as they are needed. Constitutive enzymes are always present in a cell. Examples are genes for most of the enzymes in glycolysis. For these genetic regulatory mechanisms, the control is aimed at mRNA synthesis.

## The Operon Model of Gene Expression

In bacteria, a group of coordinately regulated structural genes with related metabolic functions and the promoter and operator sites that control transcription are called an operon. It is also defined as a cluster of genes whose expression is controlled by a single operator. An operon is essentially consisting of following components:

1. *Promoter* : A site on the DNA where the RNA polymerase binds and begin the protein synthesis
2. *Operator* : A site on the DNA where the repressor protein binds and blocks the mRNA synthesis
3. *Repressor protein* : A protein bind in the operator region and block the mRNA synthesis
4. Structural genes: They are the actual coding region of the proteins or enzymes. This portion of the gene is also referred as open reading frame, because this portion alone is read by RNA polymerase to transcript the mRNA.
5. *Terminator* : This portion of DNA of the operon will terminate the mRNA synthesis by forming loop like arrangement in the same strand
6. *Ribosome binding site* : some 6 base pairs next to promoter, where ribosome bind with mRNA during translation process

These are the components of any operon. As a model, the diagrammatic representation of lactose utilizing genes, referred as *lac* operon was given in the below diagram.

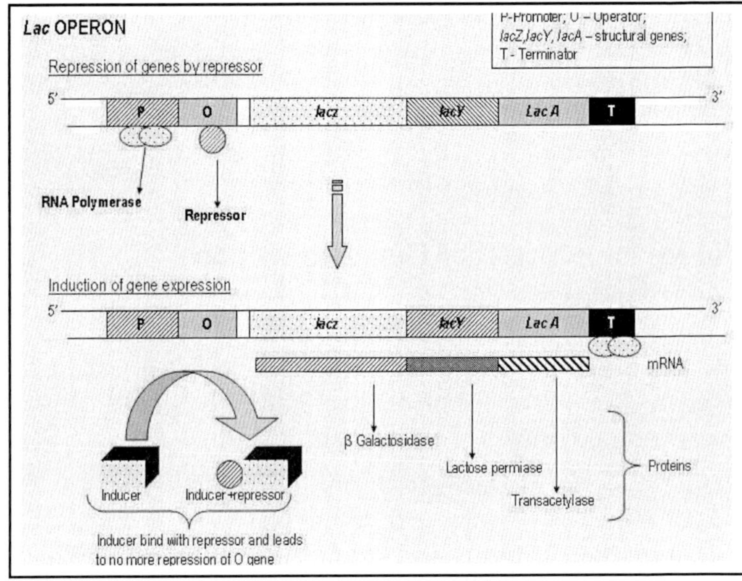


*Lac operon – suppression & Expression*

P denotes the promoter region, in which the RNA polymerase binds; O denotes the operator region, in which the suppressor protein binds and blocks; *LacZ, LacY* and *LacA* are the structural genes which code the enzymes namely, β galactosidase, Lactose permiase and lactose transactylase (which are essential for lactose utilization); T denotes the terminator which stops the mRNA synthesis.

Under normal condition, meaning, no lactose in the medium, the repressor blocks the operator of the lac operon and thereby no production of these enzymes.

When lactose was introduced to medium transport to inside of the cell, act as *inducer*. The inducer will bind with repressor protein and lead to no more blockage of operator region. Then RNA polymerase starts synthesizing mRNA up to the terminator region. The transcription will happen until the presence of inducer (lactose). When lactose was exhausted in the medium, the repressor protein will again block the operator and no more mRNA synthesis take place.

This kind of gene regulation is referred as *repression* and some times induction type gene regulations are also common one.

## Importance of Microbial Genetics

1. To understand the gene functions of microorganisms.
2. Microbes provide relatively simple system for studying genetic phenomenon and thus useful for other higher organism.
3. Microorganisms are used for isolation and multiplication of specific genes of higher organisms (which is referred as gene cloning).
4. Microbes provide many value added products like antibiotics, growth hormones etc. Microbial genetics will be helpful to increase these products' productivity by microbial biotechnology.
5. Understanding the genetics of disease causing microorganisms especially virus, will be useful to control the diseases.
6. Gene transfer among the prokaryotes play major role in the spread of the genes in a particular environment. Microbial genetics will be useful to study the gene transfer form one organism to another.

## Mutation

Is an inherited change in the base sequence of the nucleic acid comprising the genome of an organisms. A strain carrying such change is called as *mutant* strain and non-mutated which was isolated form natural environment is called as *wild type* strain.

Genotype of an organism for particular character can be written as three lettered *lower case with italic letters* following a capital letter. Ex. *lacZ* refers the gene which codes the enzyme β-galactosidase. The protein transcripted from this can be written as normal letters starting with capital letter. For the above gene, the protein may be written as LacZ. The phenotype of the organism for this gene (whether present or not) can be written as LacZ $^{+}$ for presence and LacZ $^{-}$ for absence.

## Mutant identification

In a wild type organism very rarely the mutation will occur with very low mutation frequency. (may be 1 in $10^6$ to $10^8$ cells). So, identifying the mutant from the wild type is a key phenomenon.

The mutants can be identified from a population either by *screening* or by *selection*. Screening refers the procedure that permit the sorting of organism by phenotype or genotype. Selection refers the procedure in which pressure will be given to grow a particular genotype.

For example, in a mixed population, a particular mutant lost its colour (pigmentation), we have to go for screening procedure to identify the mutant. In this case, we cannot use selection to identify the mutant. This type of mutant identification is very laborious and limited one.

On the other hand, in a mixed population, a particular mutant gain an antibiotic resistance, (the wild type may be susceptible), we can use selection procedure to identify the mutant by growing the mixed population over the medium incorporated with the particular antibiotic, which will allow mutant alone to grow on that. This type of mutant identification (selection) is a powerful tool to identify the mutant.

## Replica Plating Technique

This is also a selection type of mutant identification procedure which will be useful to identify *auxotroph* mutants. Auxotroph mutant is a metabolic defective mutant. The mutant which cannot synthesize a particular nutrient, required the nutrient through externally is referred as *auxotroph* mutant. The strains which can synthesize the nutrients by themselves are referred as *prototrophs*. The auxotrophic mutants are very common for amino acids. Normally the wild strains can able to synthesize all the essential amino acids by themselves. By mutation, if a strain lost its ability to synthesize particular amino acid (Ex. Leucine), the mutant is referred as *leu*⁻ auxotroph mutant. This mutant can grow only the medium supplemented with leucine, whereas the prototroph can grow in the medium without leucine. This type of auxotrophic mutant can be selected by replica plating technique.

The procedure of replica plating technique is as follows: The master plate (mixture of prototroph and auxotroph) will be duplicated to medium with and without leucine by means of rubber stamp like device (called as replica plate). A clone which grows in medium with leucine and which cannot show growth on medium without leucine is an auxotrph.

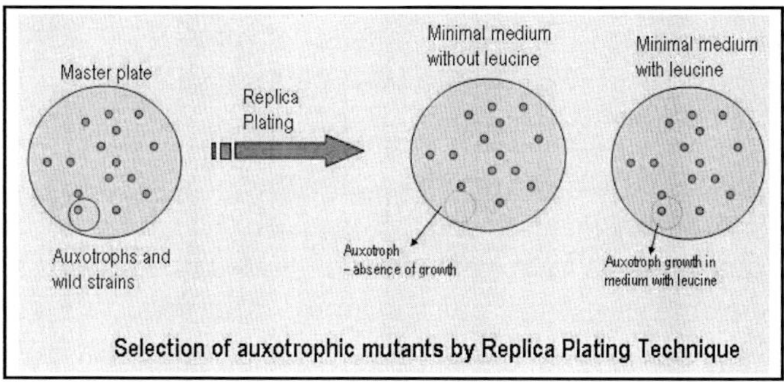

Selection of auxotrophic mutants by Replica Plating Technique

## Spontaneous & Induced mutation

The natural radiation (cosmic rays, UV rays etc) alter the genotype of an organism which is called as spontaneous mutation. Induction of mutation by some agents artificially is referred as induced mutation.

## Point Mutation

Mutation occurs in one or two base pairs of a gene is referred as point mutation. The other mutation is change in the segment of chromosome, referred as chromosomal aberration. Since bacteria are having single filamentous double strand DNA as their genome, the point mutation is a common one.

The point mutation occurs by means of two ways as A.Base pair substitution B. Frame shift mutation.

*A. Base Pair substitution*

One nucleotide base pair is replaced by another base pair, such mutation is referred as base pair substitution. One purine (Adenine or Guanine) is replaced by another purine or one pyrimidine (Thiamine or cytosine) is replaced by another pyrimidine is called as *transition.* One purine is replaced by pyrimidine or one pyrimidine is replaced by purine is called as *transversion.*

The base pair substitution type mutation alters the base pair arrangement of a gene and leads to alter in the transcription (mRNA) and translation (protein). There are three types of effects due to base pair substitution.

1. *Silent mutation* : The change of base pair, that codon codes the same amino acid (as that of wild type), such mutation is silent mutation. (Because there are four codons are available for single amino acid). In this, mutation occurs, but the amino acid sequence of the protein is not changed.
2. *Nonsense mutation* : The change of base pair, that codon stops the protein synthesis (due to change as stop codon), the effect is referred as non sense mutation. This base pair substitution will stop the protein synthesis which yields the incomplete protein.
3. *Missense mutation* : The change of base pair, that codon codes some different amino acid; such effect is referred as missense mutation. Here, the protein will be synthesized but with altered amino acid sequence.

The diagrammatic representation of these effects is provided with an example as follows:

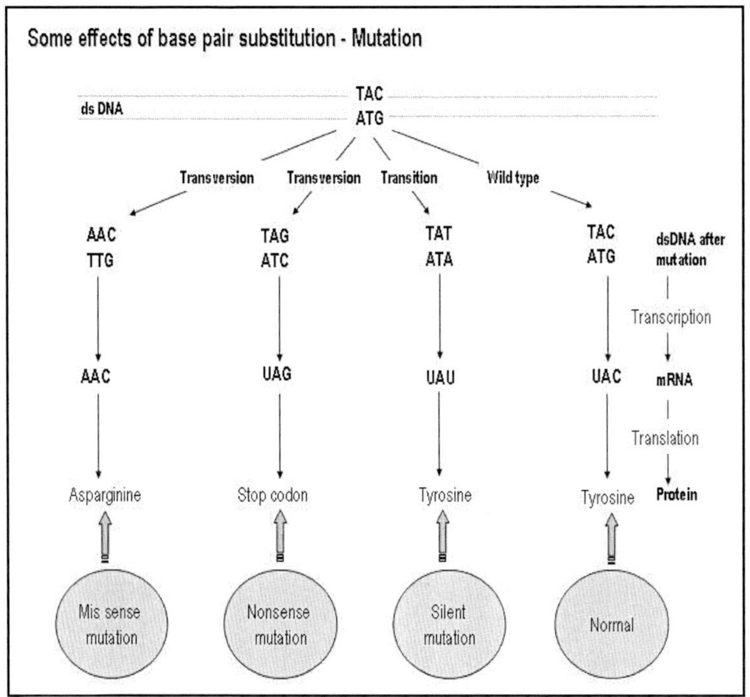


*B. Frame Shift Mutation*

A type of point mutation in which mutation occurs by addition or deletion of a base pair , which is referred as frame shift mutation.

Addition or deletion of base pair leads to change or shift the reading frame during transcription resulted in change in the amino acid sequence of protein.

Addition of base pair is referred as addition mutation and deletion of base pair is deletion mutation.

The diagrammatic representation of these effects (addition or deletion) is provided with an example as follows:

**Some effects of Frame Shift Mutation**

| | DNA | mRNA | Amino acid sequence |
|---|---|---|---|
| Normal | ATG CCT GTT ATG A<br>TAC GGA CAA TAC T | AUG CCU GUU AUG A | Met-Pro-Val-Met- |
| Deletion (deleted) | ATG CCT GTT ATG A<br>TAC GGA CAA TAC T | AUG CCG UUA UGA | Met-Pro-Leu-STOP |
| Addition (Added) | ATG CCT A GTT ATG A<br>TAC GGA T CAA TAC T | AUG CCU AGU UAU GA | Met-Pro-Ser-Try- |

## Reversion

The mutation which discussed above is reversible, that means the mutants can revert to the wild type. This phenomenon is referred as reversion mutation and the strain is called as revertant.

A second site mutation which restores the wild type genotype is referred as *suppression mutation.*

## Insertion sequences

About 700 – 1400 base pair sized DNA fragments ( as identified genes) inserted in the wild genotypes, cause mutation are referred as insertion sequences. The gene responsible for such character of insertion sequence is *transposase gene.*

## Phenotypic variations observed in microbial mutations

The following are some common phenotypic characters observed during mutation.

| Phenotypes | Cause |
|---|---|
| Auxotrophs | Loss of enzymes to biosynthesis a particular nutrient (especially amino acids) |
| Cold sensitivity | Alter the cold resistant protein |
| Drug resistance | Increase the permeability and detoxification of drugs |
| Non-capsulate | Loss of capsules |
| Loss of pigments | Loss of enzymes for biosynthesis of pigments |
| Rough colony | Loss of polysaccharides particularly extracellular, which lead to dry and rough unshaped colonies |
| Sugar fermentations | Loss of fermentative path way enzymes |
| Non-motile | Loss or unfunctional flagella |
| Temperature sensitivity | Alter the heat resistance proteins |
| Virus resistance | Loss of viral receptors |

## Mutagens

The chemical, physical and biological agents that can induce the mutation are referred as mutagens. The following are some of the features of mutagens.

### 1. Chemical mutagens

The chemicals can be divided into four groups based on their action namely.

- base analogs - Incorporate in the DNA instead of particular N base;
- DNA reacting chemicals - react with DNA nucleotide (especially sugar);
- Alkylating agents - change the structure of DNA and.
- Intercalating dyes - these chemicals will insert between base pairs of DNA.

| S. No. | Chemical | Action on base pair of DNA | Result |
|---|---|---|---|
| **a. Base analogs** | | | |
| 1. | 5-bromouracil | Incorporated instead of T | AT → GC |
| 2. | 2-aminopurine | Incorporated like A | AT → GC |
| **b. Chemicals react with DNA** | | | |
| 1. | Nitrous oxide | Deaminates A and C | AT → GC, GC → AT |
| 2. | Hydroxylamine | Reacts with C | GC → AT |
| **c. Alkylating agents** | | | |
| 1. | Ethyl methane sulfonate (EMS) | Add methyl group on G lead to faulty pairing with T | GC → AT |
| 2. | Nitrosoguanidine | Cross link the DNA strands | Deletion mutation |
| **d. Intercalating agents** | | | |
| 1. | Acrydine orange, Ethydium bromide | Insert between base pairs | Addition mutation |

## 2. Physical agents

Both non-ionizing and ionizing radiations able to cause mutation. Non-ionizing radiation has wider use in the microbial mutations.

- *UV light* : This non-ionizing radiation can be absorbed by purine and pyrimidine strongly at 260 nm. Two adjacent pyrimidines form a *dimer*, by which covalently joint. During replication, the DNA polymerase wrongly recognize the pyrimidine and add a wrong base pair which leads to mutation.
- *Gamma rays* : This ionizing radiation at low concentrations react as free radical attack and DNA breakage which leads to mutation. Gamma rays are not commonly used for microbial mutations, instead, highly useful for mutating plant materials.

## 3. Biological agents

Transposable element, insertion sequences and Mu phage are some of the biological agents cause mutation.

- T*ransposable element* : A genetic element that can move from one place of chromosomal DNA to another place. Such element's insertion lead to mutation which is referred as transposon mutagenesis.
- *Insertion sequences* : The simplest type of transposable sequence (about 700 - 1000 bp) which has the transposase gene alone, which causes the mutation.
- *Mu phage* : This double strand DNA containing temperate bacteriophage can able to act as transposable elements, causes mutation by transposition.

## Site directed Mutation

So for we have discussed about the mutations which occur at random, that means, the gene which is mutated and type of mutation cannot be predicted, referred as random mutation.

The present modern molecular techniques allow to mutate an organism at specific portion of gene by means of special vectors (plasmids) with controlled system. This type of mutation is referred as site directed mutation.

## Genetic Recombination

Genetic recombination involves the physical exchange of genetic material between organisms. It involves the genetic exchange between homologous DNA sequences form two different organisms. In classical genetics, it is referred as crossing over. In prokaryotes, *RecA protein* is a DNA binding protein involved in the genetic recombination. The following sequence of steps involved in the genetic recombination of prokaryotes.

The cell which gives DNA is *donor* and which receives the DNA is *recipient.* In prokaryotes, the genetic recombination is observed because of homologous DNA from one cell to another by three processes.

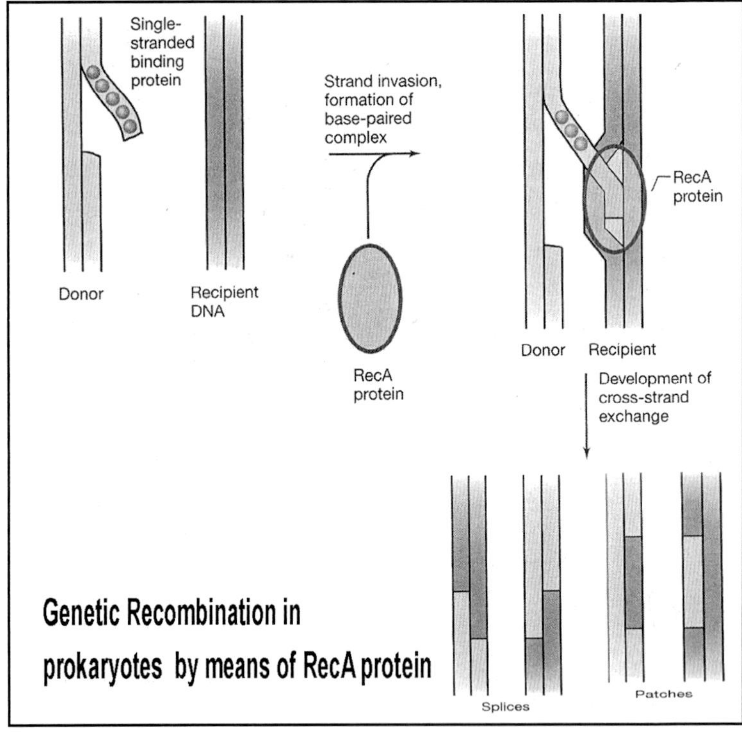


**Genetic Recombination in prokaryotes by means of RecA protein**

1. *Transformation*: Donor DNA free in the environment (naked DNA) transferred to recipient cell.
2. *Transduction*: Transfer of donor DNA to recipient cell by means of virus.
3. *Conjugation*: Transfer of donor DNA to recipient cell by cell to cell contact.

## 1. Transformation

A naked DNA or cell free DNA incorporated into the recipient cells and brings about genetic recombination is referred as transformation.

Most of the bacteria both G+ and G- are easily transformable, but only a fragment of DNA can easily be transformed (few number of genes).

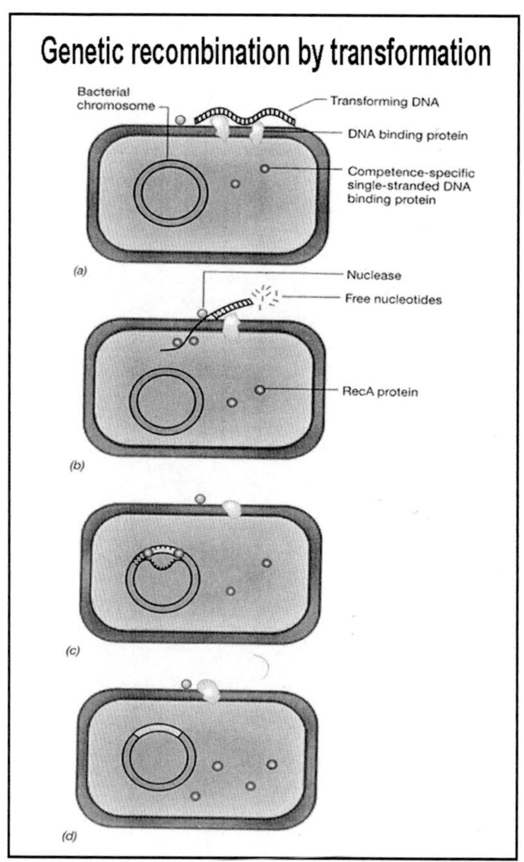


The ability of an organism to take up the DNA molecule is referred as *Competence* and the organism is said to be *Competent.* The competence of organisms varies among the organisms. *Bacillus* has 20% competence cells and *E. coli* and *Streptococcus,* 100% of cell become competence.

The gram-ve bacterium, *Haemophilus* takes only double strand DNA whereas, the *Streptococcus* and *Bacillus* can take only single strand DNA , while the complementary simultaneously degraded.

The naked DNA first bound on the surface of the recipient cell with the help of surface proteins. A single strand of DNA is allowed to enter inside the cell. The entered DNA will be moved to chromosomal DNA and genetic recombination takes place with the help of RecA protein.

### Artificial transformation

The molecular biology techniques now used to transfer any naked DNA (in the form of vectors) to recipient cells are artificially.

1. *Calcium chloride mediated transformation* : The $CaCl_2$ washed recipient cells mixed with DNA allowed a temperature shock (42°C for 40-60 sec.) will allow the naked DNA to get transformed.
2. *Gene gun (Particle gun)* : will allow transformation of plant tissues, yeast, algae etc.
3. *Electroporation* : Transformation takes place by means of electrical pulses.

## 2. Transduction

A DNA transferred from one cell to another through the agency of virus is referred as transduction. The transduction of bacteria occurs through bacteriophages.

Some phages are lytic, which kill the host cell during replication, called *lytic cycle* and *lytic phages*. Ex. T2 phage; Some phages replicate in the bacteria without killing the host cell, called as *lysogenic cycle* and *temperate phage*. The host cell is called as *lysogen*.

Based on the behavior of virus, transduction occurs in two way. 1. Generalized transduction and 2.Specialized transduction.

### Generalized transduction

The generalized transduction occurs due to lytic phage. Normally, the lytic phage send its DNA after adsorption, into host cell and DNA will order the chromosomal DNA to synthesize viral DNA and viral proteins. Then packing take place and virions lyse the host cell and escape to the environment. Accidentally, sometimes, the viral protein packed with bacterial DNA instead of viral DNA during lytic cycle, are referred as *transducing particle*. This transducing particle can able to *infect* the host cell, but cannot perform the normal lytic cycle. When the transducing particle infect the recipient cell, the DNA will penetrate into cytoplasm and by the help of RecA protein get into chromosomal DNA, which leads to genetic recombination (As shown in the diagram).

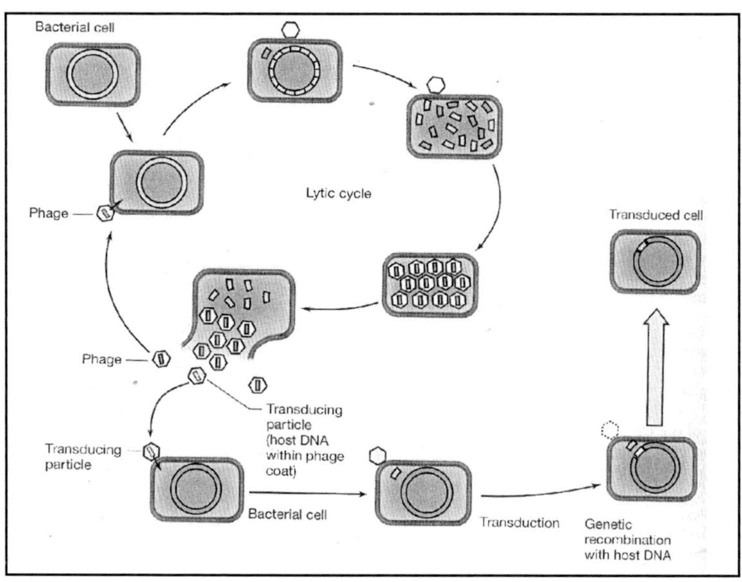


The frequency of transduction is 1 in $10^6$ to $10^8$ cells. This method of transduction is generalized one. Any portion of donor DNA can become transducing particle and become transduced transduction takes place; this type of transduction is referred as *generalized transduction*.

*Specialized transduction* : On the other hand, some temperate phages can able to take up a specific portion of donor DNA as their part and transduction takes place, which is referred as *specialized transduction*.

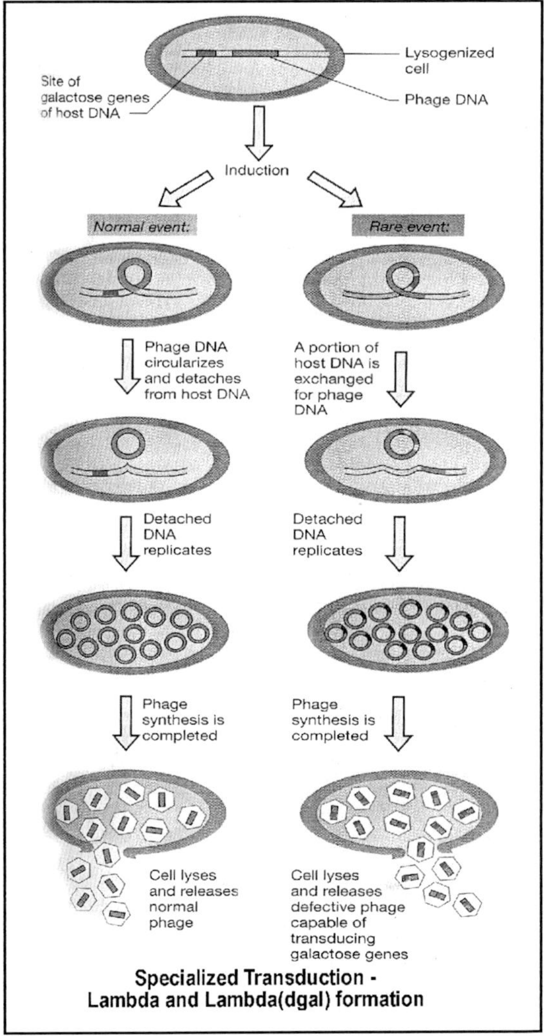


**Specialized Transduction - Lambda and Lambda(dgal) formation**

The temperate phage (Lambda phage) send its DNA after adsorption. The penetrated viral DNA will get integrated into bacterial chromosomal DNA and persist for ever. This stage of phage is referred as *prophage*. The cycle is lysogeneic cycle. Due to chemical or physical induction, the prophage get activated, replicated its viral DNA and protein coat followed by assembly and lysis of the cell.

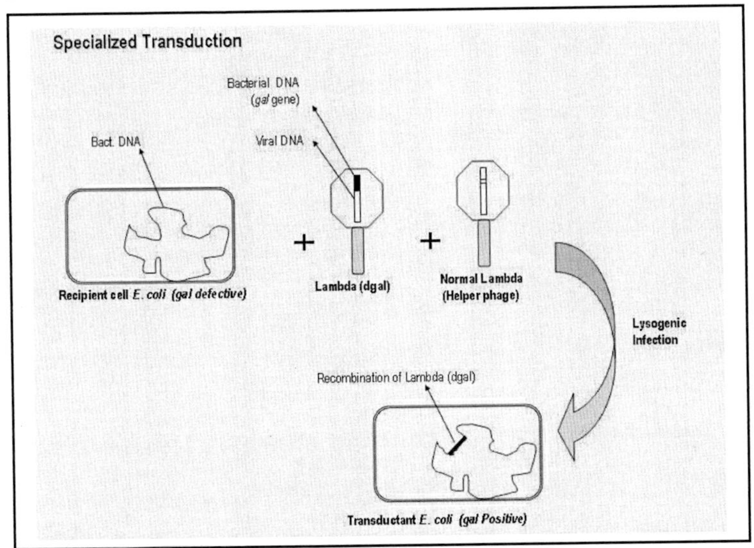


The lambda phage will get integrated into bacterial chromosome at specific region (near to *gal* gene - galactose operon). When induce the prophage, sometimes, the viral DNA wrongly included some portion of bacterial DNA (mostly *gal* gene) with loss of portion of viral DNA. After assembly, these phages will escape to environment. These phages are designated as lambda (*dgal*) phages. These λ(*dgal*) phages cannot act as that of normal lambda phages. For infection of recipient cells, the λ(*dgal*) phages need the help of normal lambda phages (referred as helper phages). When these λ(*dgal*) phages infect the *E. coli* (gal defective *gal*$^-$), the transduction take place lead to gain of *gal* gene. The recipient cells become gal$^+$.

### 3. Conjugation

Bacterial conjugation (mating) is a process of genetic recombination that involves cell to cell contact.

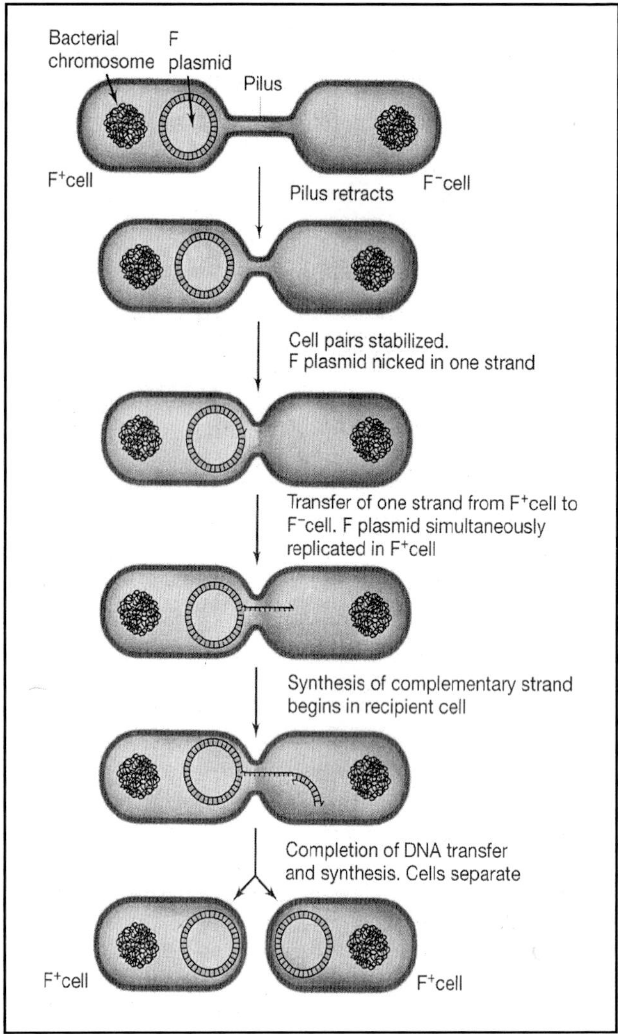


Genetic Recombination by Conjugation

The gene for conjugation (ability to mate) is a plasmid borne character. The gene involves the conjugation is designated as F factor (fertility factor) which located in the plasmid (referred as F plasmid). The cell which has the F plasmid is donor (male), designated as $F^+$ cells and cell doesn't have the F plasmid is the recipient (female), designated as $F^-$ cells. The DNA will be transferred from donor to recipient through physical contact. So another requirement for cell to cell contact is *sex pilus* through which the donor DNA will be transferred to recipient cell.

The diagrammatic representation shows how the conjugation takes place. The sex pili makes a connection between donor and recipient. A nick (split) in the complementary strand of F plasmid takes place. The single strand plasmid transferred to $F^-$ cell and become double stranded. The $F^+$ cell also synthesized its cDNA and become double strand. After conjugation, both the cells become $F^+$ cells.

## Hfr cells

Sometimes, F plasmid transferred from $F^+$ cell to $F^-$ cell get integrated into recipient cell chromosomal DNA. Such conjugants possess the F factor in the chromosomal DNA instead of plasmids. They are referred as *hfr cells* (High Frequency Recombination). They have very high frequency of conjugations.

## Other genetic recombinations

The process by which a gene moves from one place of chromosome to another in a genome is called as transposition. The genes involved are transposon genes or transposable elements. Insertion sequences are also a type of transposable element (of about 700 bp size) which also can cause the genetic recombinations. These DNA are used to insert a foreign DNA to an organism and also cause mutagenesis (discussed earlier).

## Genetic Engineering

Development of sophisticated procedures for isolation, manipulation and expression of genetic materials in organism, a field is called as *genetic engineering*. Isolation, purification and replication of fragment of DNA through some vectors are referred as *gene cloning*. The plasmids and bacteriophages are used as vector for gene cloning.

CHAPTER

13

# Industrial Microbiology

Industrial Microbiology is the discipline that uses micro organisms, usually grown on large scale, to produce commercial products or carry out important chemical transformations.

The products derived from such processes are

- Alcohols and alcoholic beverages.
- Production of pharmaceutical agents like antibiotics, vaccines etc.
- Food additives
- Enzymes
- Chemicals
- Single cell proteins

All these industrial microbiological processes are enhancement of metabolic reactions of microorganisms with a goal of over production of these products.

Now the microbial biotechnology joined with the industrial microbiology, by means of genetic manipulations of these microorganisms lead to many fold increase of products and also to produce new products, which are originally not produced by these microorganisms.

Products of microorganisms can be grouped into three, based on their activity.

- *As cell* : The microbial cell itself as a product of the industrial process. Ex. Baker yeast, Single cell proteins like yeast, *Chlorella* etc.
- *Bioconversion* : Microorganisms act as a biocatalyst to convert one substrate to another value added product. Ex. Steroids
- *Products from cell* : The products will be separated from microorganisms after the large scale production is over. Ex. Alcohols, antibiotics, enzymes etc.

## Fermentation

In industrial term, fermentation refers to any large scale microbial process carried out either aerobically or anaerobically.

## Product formation

The product formation during fermentation (or microbial growth) was of two groups.

1. *Primary metabolites* : The product formed during growth of an organism. It means that the product will be formed during the active growth stage (log phase) of the microorganism. Ex. Alcohol by yeast. Since alcohol is produced by the primary metabolic pathway of yeast (EMP pathway), the alcohol will be accumulated during the active growth stage of yeast.
2. *Secondary metabolites* : The products formed at the end of active growth stage or at the beginning of the stationary phase are referred as secondary metabolites. Ex. Antibiotics are produced usually at the end of log phase. These products were formed by the secondary metabolic pathways which are not essential for its growth. These kinds of metabolites are produced only by few organisms.

By altering the conditions and strain improvement, it is possible to get over production of secondary products, whereas it is difficult to get over production of primary metabolites.

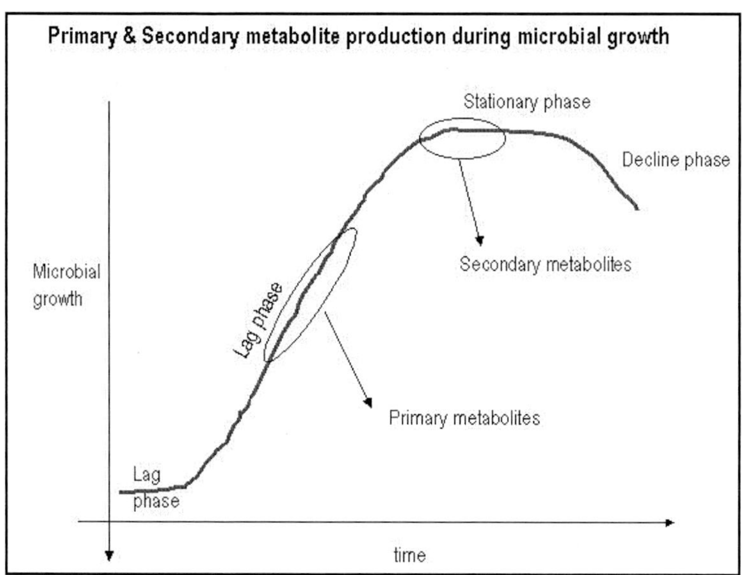


## Types of Fermentation

Based on the state, the fermentation can be classified in to two.

*Submerged (liquid) fermentation* : Fermentation is carried out as liquid forms. Ex. Alcohol from sugarcane industrial waste – Molasses by yeast

*Solid fermentation* : Fermentation is carried out as solid condition with about 30-50 % moisture level. Ex. Gibberelic acid production from grains by fungus *Fusarium moniliformi*

Based on the microbial growth (culturing method):

- *Batch fermentation* : A closed system of microbial fermentation in which the volume is fixed.

- *Fed batch fermentation* : The substrate for fermentation (raw material) is added in increments as the fermentation progresses.
- *Continuous fermentation* : An opened system of fermentation in which the input (nutrients solution) is added continuously with a withdrawal of equal portion of converted or fermented solution.

## Large-scale fermentations

The vessel in which industrial fermentation process carried out is called as fermentor. The volume of the fermentor varies from 5-10 lit to 500,000 lit. Similarly, the fermentor model will vary for aerobic process and for anaerobic process.

The basic structure of fermentor is as follows:

The components are

- *Inlet* : to feed the substrates, microorganism etc
- *Outlet* : to collect the products, to take samples
- *Stirrer* : to mix the contents so as to get contact between nutrients and microbes
- *Air sparger* : to provide oxygen to the microbes by small air bubbles
- *pH meter/controller* : to check the pH and if needed to add acid and alkali to maintain optimum pH
- *Thermometer* : to check the temperature
- *Outer jacket* : by which cool water will be circulated to cool the fermentor (because during fermentation, lot of heat will be generated)
- *Sterilization unit* : by which steam will be supplied to the fermentor to sterilize the medium and container

*Scale up* : Scale up refers the step by step process in any fermentation to get the maximum product.

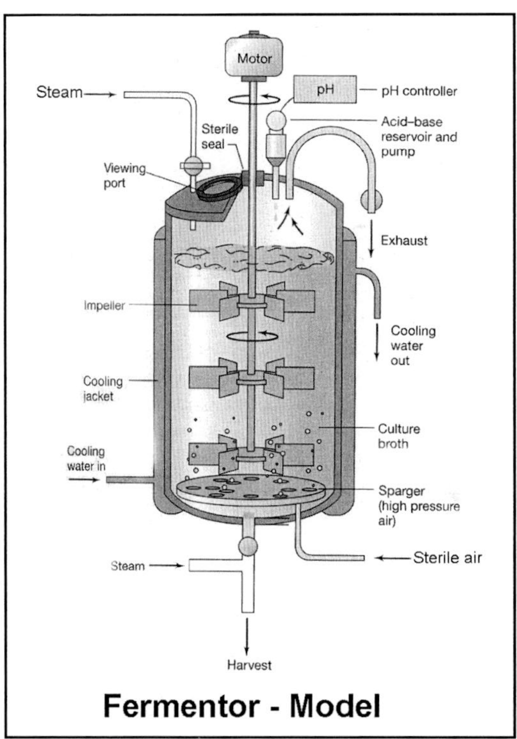


**Fermentor - Model**

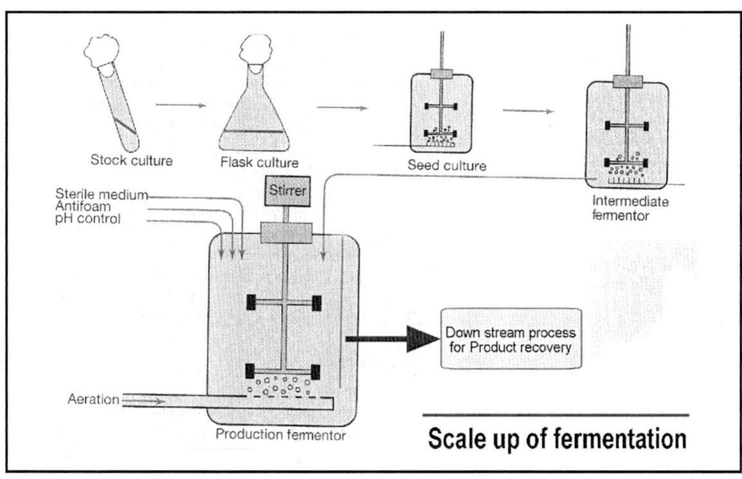


**Scale up of fermentation**

## Products from Industrial Microbial Fermentations

The following list show the commercial products obtained through fermentation by different micro-organisms.

### a. *Antibiotics*

Most of the antibiotics are produced directly though microbial fermentations. Sometimes, the antibiotic produced by an organism will be again modified through chemical process, referred as semi-synthetic antibiotics. The table shows some antibiotics and the organisms involved in commercial production.

| Antibiotics | Organism |
|---|---|
| Bacitracin | *Bacillus licheniformis* |
| Cephalosporin | *Cephalosporium* |
| Streptomycin | *Sreptomyces griseus* |
| Kanamycin | *S. kanamycetus* |
| Erythromycin | *S. erytheus* |
| Polymyxin | *Bacillus polymyxa* |
| Penicillin | *Penicillium chrysogenum* |

### b. *Vitamins*

Vitamin B12 and riboflavin are commercially produced by microorganisms. *Propionibacterium* and *Pseudomonas denitrificans* produce vitamin B12 and fungi *Ashbya gossipii* produce riboflavin.

### c. *Amino acids*

L-glutamate, L-aspartate, glycine, cysteine, tryptophan, aspartate, monosodium glutamate (MSG) are common amino acids commercially produced by microbes. Presently, the genetically modified bacteria are used to produce these amino acids.

d. *Bioconversion*

In bioconversion process, the microorganisms act as biocatalyst for conversion of one compound to another product. Ex. Most of steroids are produced by bioconversion process.

e. *Enzymes*

The enzymes are mostly used for food industry and molecular biology. The following table shows the enzymes and responsible organisms.

| Enzyme | Organism |
|---|---|
| Amylase | *Bacillus licheniformis, Aspergillus* |
| Protease | *Bacillus licheniformis* |
| Pectinase | *Aspergillus niger* |
| Glucose isomerase | *Bacillus coagulans* |
| Rennin | *Streptococcus lactis* |
| Lipase | *Aspergillus, Candida* |
| DNA polymerase | *Thermus aquaticus* |

f. *Chemicals*

Vinegar (acetic acid), citric acid, acetone, butanol etc. are some examples of the chemicals produced by microorganisms.

Vinegar - *Acetobacter, Gluconobacter;* Lactic acid - *Lactobacillus;* Citric acid - *Aspergillus niger,* Acetone - *Clostridium acetobutylicum.*

g. *Single cell protein*

The proteins produced by microorganism as food is referred as single cell proteins (SCP). The yeast, *Chlorella,* cyanobacteria, *Spirulina* are some common examples of SCPs.

h. *Alcohols and alcoholic beverages*

The yeast play a major role in the alcohol industry. Since, it is a facultative aerobe, can be grown aerobically and anaerobically. When, yeast grown aerobically, its biomass (cell yield) increased with low alcohol, used for production of baker yeast. Whereas, yeast grown anaerobically, the alcohol production will be maximum.

*Brewing* refers the manufacture of alcoholic beverages from fermentation of malted grains. Wine is obtained form grape; beer from barley; whiskey, rum, otka, etc are the commercial preparations of alcohols. The beer is differentiated form other beverages by its processing method. After fermentation, the beer is prepared as such without distillation, instead, pasteurization and packing gives beer. Whereas other commercial beverages are prepared after distillation of alcohol.

i. *Food*

The microorganisms as such can be used as food and by their action also many foods are produced. Following are some of the foods obtained form microbes.

- *Mushroom* : *Agaricus bisflorus* (button mushroom); *Pleurotus sajor-caju* (oyster mushroom).
- *Bread* : by yeast
- *Butter, yogurt, like dairy products* : Lactic acid bacteria (*Lactobacillus, Leuconostoc*).

j. *Vaccines*

The vaccines are the immunized antigens which has whole cell or part of cell or cell product. The attenuated virus is a best example for whole cell virus vaccine; Presently the viral DNA also act as vaccine which is referred as DNA vaccines. All these are commercially produced.

CHAPTER

14

# Immunology — Host Defence Mechanism

The pathogenicity of an organism varies form mild attack to severe fatal attacks. The host has many mechanisms to resist or to avoid the pathogens apart from immune systems. One or few of these will be activated when the pathogen enters into the body. Following are some of the listed natural host defense mechanisms.

1. *Natural host resistance:* The host has the natural ability to resist the pathogen. For example, the anthrax may cause mild pustules to man whereas cause fatal blood poisoning to cattles.
2. *Age, stress and diet:* Very young and very old are susceptible whereas the adults are generally resistant against pathogens. Similarly the physical stress induces the susceptibility of the host against pathogens. Malnutrition is also allowed the pathogens to act very rapidly.
3. *Anatomical defenses:* The body of host (man) has so many defense mechanisms to avoid the pathogens. lysozyme of tears; rapid passage of air in nose; highly acidic pH of stomach ($pH_2$); rapid change in pH of intestinal track; native microflora of large intestines; skin layers; mucous membranes of lungs; coughing; sneezing are some anatomical defenses of our body to resist the pathogens.

4. *Fever and inflammation:* Redness, swelling, pain, heat and fever are some non-specific reactions carried out by host against pathogens.
5. *Immune system:* The specific and non-specific immunity is the well-advanced resistance mechanism, which will be discussed in this chapter.

## Immunology

Immunology is the study deals with the mechanism by which an organism able to resist the infection by pathogens.

Immunity is defined as the ability of an organism to resist the pathogens' infection. The immunity can be classified in to two groups.

1. *Non-specific immunity* – refers the general ability of certain cells of the body to resist the most of the pathogenic organisms like bacteria, fungi, virus etc.
2. *Specific immunity* – refers highly sophisticated mechanism found in the vertebrates for developing resistance to individual pathogens.

These two immune systems were triggered when the foreign molecule (pathogenic microorganisms or their toxins) enters to the cell. These foreign molecules, macromolecule components of pathogens, such as surface proteins and excreted proteins (toxins) are collectively referred as *immunogens or antigens*.

In response to antigen, the host immune system will produce soluble proteins to detoxify the antigen's effect is referred as *antibody*.

Once antibody is produced by an immune system for particular antigen, the second time infection by same microorganism or same antigen will be resisted by rapid antibody production. This capacity of immune system for responding the challenge of additional exposure to pathogen or antigen is referred as *immune memory*.

## Blood and Lymph

Blood and lymph are the two major body fluids responsible for all these immune systems.

Blood consists of both cellular and acellular components.

*Cellular components*

- *Erythrocytes* (red blood cells) – Non nucleated cells; major function is the $O_2$ carrier to cells
- *Leucocytes* (white blood cells) - involves in the immunity; two kinds are present
- *Phagocytes / monocytes :* They are the phagocytic cells (eating cells) which will engulf the pathogens
- *Lymphocytes:* responsible for antibody production against antigens

*Acellular components*

- *Plasma* – When the cellular components removed from blood, is referred as plasma. It has an important protein called fibrinogen, which is responsible for blood clotting.
- In plasma, if the blood clotting protein is removed, that fluid is referred as *Serum.*

*Lymph:* is a fluid similar to blood but lack of red blood cells and travels through separate circulatory system (the lymphatic system). This system has capillary like arrangement called lymph node to filter the pathogenic cells.

## Antigens or Immunogens

They are the substances when admitted to host (vertebrates) able to induce the immune system of the body, especially specific immune system, are referred as antigens or immunogens. These include variety of macromolecules of pathogens namely many polysaccharides, surface proteins, nucleic acids, lipoproteins etc.

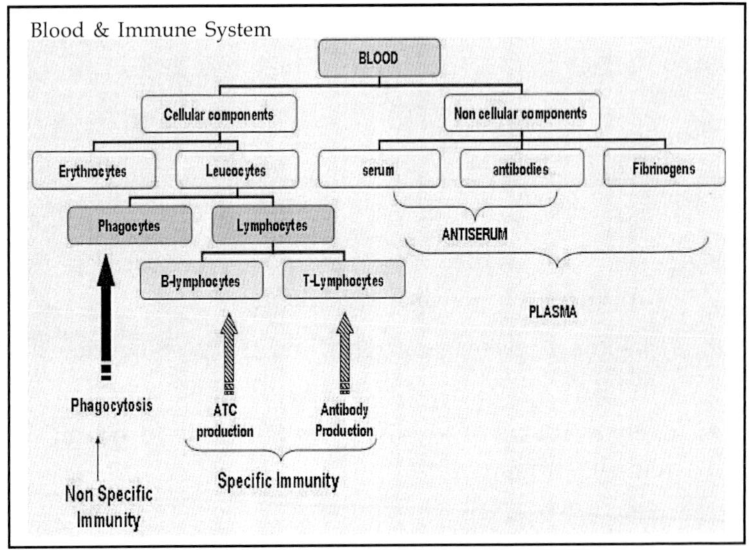


- Some times, the organism especially bacteria secrete protein extracellularly, which produce immediate damage to the host cell are referred as *exotoxins*. Such exotoxins released at small intestine of host are referred as *enterotoxins*.
- Sometimes, the bacterial cell wall lipopolysaccharides act as toxin when solubilized, which is referred as *endotoxins*. - All these are act as antigens.

## Non-specific Immunity

refers the general ability of *certain cells* of the body to resist the most of the pathogenic organisms like bacteria, fungi, virus etc. The non-specific immunity is a common resistance for all pathogens where as the specific immunity is specific to every pathogens. The common non-specific immunity is phagocytosis. (engulfing the pathogens).

## Phagocytes and Phagocytosis

Some leucocytes (white blood cells) found in the cell are phagocytes. They are motile, amoeboid action, which contain lysosomes. By chemical attraction, the phagocytes move towards pathogens and engulf them. By the action of lytic enzymes, they kill the pathogens. This phenomenon is referred as phagocytosis.

There are two groups of phagocytes in blood.

- *Neutrophiles* – short lived (2-3 days), small sized, large numbered phagocytes. They can move very fastly (40 mm/min).
- *Monocytes & macrophages* – Large sized phagocytes which live for long time. Macrophages are fixed in tissues and monocytes are circulating in blood.
- During phagocytosis
- Phagocytes changed their metabolism from aerobic to anaerobic form
- Formation of lactic acid drastically reduced the pH to 2
- Hydrolytic lysozymes excreted

Due to all these causes, the pathogenic organism will be killed. Sometimes bacterial pathogens show resistance against phagocytes because of the presence of capsules.

## Specific Immunity

Refers the ability of immune system to interact with specific antigens, which is referred as specific immunity. The white blood cells (Lymphocytes type of leucocytes) are responsible for such specific immunity. These lymphocytes produce soluble proteins against antigens, which can interact with antigen and neutralize the effect, referred as antibodies.

The lymphocytes are of two types based on their action and origin. They are B Lymphocytes (shortly as B cells) and T Lymphocytes (T cells). B lymphocytes developed from bone marrow cells (designated as B cells) and T lymphocytes developed from thymus (designated as T cells).

### B Lymphocytes

- They are really responsible to produce antibody against antigens and also for immune memory
- The antibody produced by B cells are referred as *immunoglobulins (Ig).*

### T Lymphocytes

- They have *Antigen Specific T Cell Receptor (ATC),* similar to antibody of B cells, which can interact with antigens.

They have multiple functions such as

- They recognize antigen and facilitate B cell to act on the antigen
- They recognize antigen and release ATC which can suppress the antigens.

### Antibodies

Immunoglobulins, ATCs and cytokines are the proteins produced by B cells and T cells respectively against antigens, which can be referred as antibodies.

*A. Immunoglobulins (IG)*

- They are the soluble proteins produced by T cells against antigens
- They are found in the serum, and the serum rich of antibodies is referred as *antiserum.*
- The immunoglobulins are the circulating antibodies

- Molecular weight ranges from 150,000 to 970,000 daltons
- Five classes of imunoglobulins are recognized *(Ig G, Ig M, Ig D, Ig E and Ig A).*
- All the immunoglobulins (Ig G,M,D,E) present both in blood and serum whereas Ig A is present in saliva.

When antigen is introduced to the body, the immunoglobulin production slowly increased to a level and again slowly decline in the body. This response is referred as primary antibody response. When second time introduced same antigen, the immunoglobulin production suddenly increased (10 – 100 time more than first time). This large raise of antibody production due to second introduction is referred as secondary antibody response.

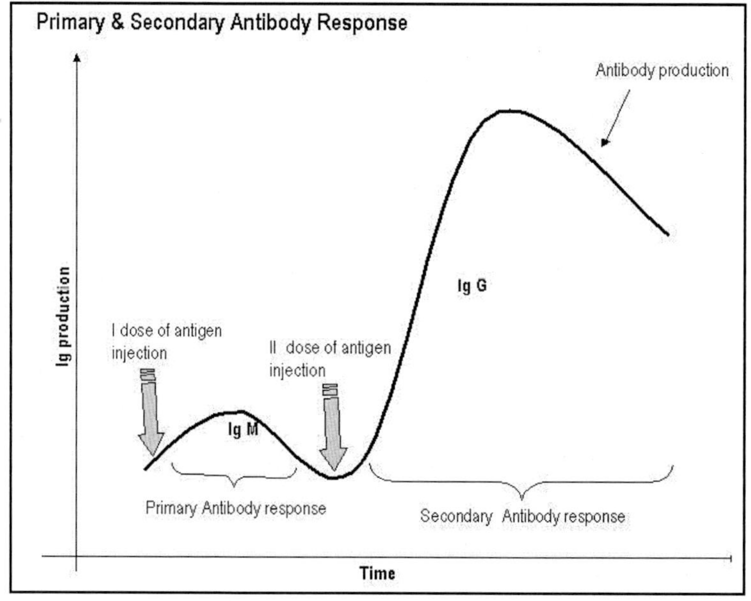


A typical immune system response is production of broad spectrum of immunoglobulin molecules against antigens. Normally a single clone of B cell will produce single type of immunoglobulin. So, When antigen is introduced, heterogeneous population of immunoglobulins will be accumulated in the serum, which is referred as *polyclonal antiserum.*

However, It is possible to produce a single antibody population (identical in structure and action) against antigens. This antibody is referred as *Monoclonal antibodies.*

- The monoclonal antibodies are used for research and clinical purpose
- Used for therapeutic applications
- Highly useful to cure cancers
- Identification of microorganisms is also possible by monoclonal antibodies.

*B. Antigen specific T cell receptors (ATCs)*

They are the surface proteins produced by T cells. They are present on the surface of the T cells, whereas the immunoglobulins are present in serum. These proteins have short chained peptides, which can bind with antigens. These proteins are referred as ATC. The T cells can interact with antigens through ATC. After a complex formed, these T cells with antigen can move to B cell or phagocytes or memory cells to deactivate the antigens.

*C. Cytokines*

A soluble immune response modulators produced by leucocytes. They are the regulatory soluble proteins produced by lymphocytes during immune response. They are responsible for signal transduction and send information inside the cell for rate of metabolism, cell division, protein synthesis etc.

Various leucocytes against antigens produce these three proteins as specific immune response.

## Antigen - Antibody Reaction

The study of antigen - antibody relation *in vitro* is called as *serology*. There are so many reactions will occur when antigen is reacted with antibody. The common reactions are as follows:

- *Neutralization* - The toxin portion of antigen may be blocked by antibody so that it cannot act to cause the disease, referred as neutralization. The antibodies are here referred as *antitoxins*.
- *Precipitation* - Generally antibodies have two binding sites. If two different antigens combined with antibody leads to precipitation of the complex which reduce the toxicity of the antigens.
- *Agglutination* - If antigen is not soluble, the antibody forms a clumping (mass like insoluble aggregates) when react with that antigen. The agglutinated clumps can be seen under microscopes.

## Immunization

So for the discussion is mainly about the immunity naturally developed against the pathogen or antigen. This immunity is inherent character, meaning the individual acquired the infection and that is controlled by natural immune system. This immune system is referred as *natural active immunity.*

On the other hand, it is possible to induce the immune system of an individual by introducing antigens to the body. This process is known as *vaccination or immunization*. It is an artificial active immunity.

The introducing material, antigen to the body is known as *vaccine*. The material used to induce the artificial immunity of the body, antigen or mixture of antigens, are referred as vaccines.

Following are some antigens used as vaccines:

- *Toxoids* – Many pathogens produce exotoxins and endotoxins. These toxins are modified and used to induce the immune system, as vaccines are referred as *toxoids*.

  Ex. *Diphtheria, Tetanus* toxoids used as vaccines.
- *Attenuated cells* – It is possible to isolate a mutant strain of pathogen which is avirulent but still has antigen are referred as attenuated strains. These cells when introduced, can induce the immunity without causing disease.

  Ex. Yellow fever, measles, mumps, polio causing viruses can be attenuated and used as vaccines.
- *Inactivated cell* – Ex. Rabies virus inactivated cells used as vaccines.
- *Killed cells* – Ex. Typhoids, Plague causing organisms - killed and used as vaccines.
- *Polysaccharide of cell* – Ex. Pneumonia causing bacterial polysaccharide used as vaccine.
- *DNA* – Ex. Hepatitis A and B viruses DNA are used after genetic recombination as vaccines.

## DNA vaccines

The specific antigen producing gene can be introduced to host cell. The gene produces antigenic protein and stimulates the immune system. This type of vaccines is DNA vaccines.

## Artificial Passive Immunity

The above discussed immunity development artificially induces the native immune system. On the other hand, introduction of antibody directly to the body is also possible. The antibodies may be produced using some other mammals like cattle, horse etc. by introducing antigen to their body.

Then, the antiserum (serum containing antibodies) obtained from cattle may be introduced to the individual. Such developed immunity is referred as *passive immunity.*

In passive immunity,

- The individual will not expose to antigens
- The individual's immune system will not be induced
- No immune memory will be there
- Immunity development is rapid.

*Note:* (All these characters are just opposite to artificial active immunity).

CHAPTER

15

# Genetic Engineering and Biotechnology

Development of sophisticated procedures for the isolation, manipulation and expression of genetic materials is a field called genetic engineering. Genetic engineering has applications in both basic and applied research. In basic research, genetic engineering techniques are used to study the mechanisms of gene replication and expression in prokaryotes, eukaryotes and viruses. For applied research, genetic engineering used for development of microbial cultures capable of producing valuable products such as human insulins, growth hormones, interferons, vaccines, enzymes etc. The commercial application of genetic engineering is sometimes referred as Biotechnology. In broad sense, Biotechnology refers the use of living organisms to carry out defined chemical processes for industrial application.

The process of isolation, purification and replication of specific DNA fragment is called molecular cloning or gene cloning. Molecular cloning is the base of most of the genetic engineering procedures. The purpose of molecular cloning is to isolate large quantity of specific genes in pure form. Thus molecular cloning is very important in genetic engineering and can be divided into several steps depends on the purpose. The steps are as follows:

1. *Source DNA/gene:* The gene of interest or total genomic DNA or synthesized from RNA template (by the enzyme Reverse transcriptase) or from DNA template (by the enzyme DNA polymerase) using Polymerase Chain Reaction (Known as PCR).
2. *Cloning vectors:* Bacterial extra chromosomal DNA, plasmids and bacteriophages are used as vectors or vehicles to replicate and carry the source DNA or gene. The DNA will be ligated with the vector and can be replicated and expressed.
3. *Host organisms:* Thus cloned vectors should be introduced to universal host organisms (usually bacterium, *E. coli*) for replication in several copies per cell and to maintain the cloned vectors.
4. Identification or screening of host clone for desired host bacterial cell with cloned vector.
5. Production of large number of desired host containing vector for isolation and study the cloned DNA.

## Gene Cloning Vectors

### *1. Plasmids and Bacteriophage Lambda*

Plasmids are the extrachromosomal DNA present in most of the bacteria, replicate independently of the host chromosome. In addition to carrying genes required for their own replication, most plasmids are nautral vectors (gene carriers) because they often carry other genes that confer important properties on their host. Plasmids have very useful properties as cloning vectors. Due to their small size, easy to isolate and manipulate, circular DNA, independent origin of replication and multiple copy number in host cells lead plasmids as ideal vectors. The plasmids pUC18, pUC19, pBR322 are some of the common plasmid vectors used for common molecular cloning.

Similarly, modified Lambda bacteriophages are highly suitable for gene cloning. Wild Lambda phage harbor too many restriction enzyme sites which is not suitable as cloning vector. The unwanted restriction enzyme sites removed by point mutation called Charon phage can be used as vehicle for insert any foreign DNA. Charon 4A, Charon 16 are some vector Lambda phages used in gene cloning.

*2. Cosmids*

Cosmids are plasmid vectors containing foreign DNA with *cos* site from Lambda genome. These *cos* sites are required for packaging DNA into Lambda virions. The main advantage of use of cosmids is to clone large fragment of DNA.

*3. Phagemids*

Phagemids are the vectors synthetically designed using basic genes from both plasmid and bacteriophages. Most of the modern vectors commercially used in gene cloning are phagemids.

## Expression Vectors

In earlier, vectors are designed mainly to clone the foreign DNA and to maintain these cloned vectors in universal host cells (commonly called cloning vectors). Later, the vectors are designed to express the foreign DNA in host cells in terms of synthesis of proteins coded by the DNA. These vectors are called expression vectors. These vectors allow the foreign genes to express in host cells, over produce the protein and to excrete them from the cell. In these vectors, the complete structural genes of the protein will be cloned in the vector with host specific promoter sequence.

## Shuttle Vectors

In advanced genetic engineering research, it will be useful to be able to move the genetic materials between completely unrelated organisms. For this purpose, shuttle vectors are used. It is one that can replicate in two different organisms. Shuttle vectors are developed that replicate in both *E. coli* and *Bacillus* and yeast, *E. coli* and mammalian cell culture like pair of organisms.

## Host for Cloning Vectors

The ideal characteristics of a host plant for cloned genes are rapid growth, capability of growth in inexpensive culture medium, not harmful or pathogenic, ability to take up the foreign DNA (competence) and stability in culture. The host must have appropriate enzymes to allow replication of the vectors. The most useful hosts for cloning are microorganisms that grow well and for which we have both sufficient genetic information and tool for genetic manipulation. These includes the bacteria, *E. coli, Bacillus subtilus* and *Saccharomyces cerevisiae*.

## Selection of Clones

A crucial step in gene cloning is finding the host that harbor the "cloned vector" (referring the plasmid with foreign DNA). For this, two kinds of selection procedures have been employed in genetic engineering.

a. *Selection:* This can allow only host containing plasmid vector by selecting for a vector marker. The most common selection marker is antibiotic resistant genes incorporated in the vector, will allow the host to grow in the medium containing antibiotics. This will allow to differentiate and select the host with vector and without vector. But, it will not differentiate the clones containing foreign inserted vectors and clones containing vectors without insert DNA.

b. *Screening:* This procedure allows to screen several clones to pick the correct clone that harbours the foreign DNA / gene inserted vector. Inactivation of β-galactosidase gene (*lacZ* gene) is the most commonly used screening procedure widely adopted to identify the cloned vector. If the *lacZ* gene containing vectors are introduced to host cells and the host cells are allowed to grow in medium containing the substance, 5-bromo-4-chloro-3 indolyl- β-D galactopyranoside (called x-gal), β-galactosidase produced from the gene *lacZ* hydrolyses the Xgal and produces the insoluble blue dye resulted blue coloured colonies in the plate. If the foreign DNA is inserted in this region (*lacZ*), then it could not able to utilize the x-gal resulted white colonies will appear. By this way (inactivation of *lacZ* gene) the screening of clones can be done to identify the foreign DNA / gene inserted vectors.

## Polymerase Chain Reaction

Polymerase chain reaction (in short PCR) is used for *in vitro* rapid amplification of DNA. PCR can multiply DNA molecules by upto a billion fold in the test tube, large amount of specific genes for cloning, sequencing or mutagenesis purpose. DNA polymerase enzyme is used in PCR reaction, which copies the DNA.

The PCR technique requires a short oligonucleotide, known as *primer*, complementary to the sequence in the gene of our interest will anneal in the source DNA (known as template DNA) and with the help of DNA dependent DNA polymerase enzyme, the complementary DNA strand will be synthesized. The steps in PCR amplification of DNA are as follows:

a. Two primers flanking the target DNA are added in great excess quantity to heat denatured target DNA (single stranded DNA).

b. As the mixture cools, the excess of primers related to the target DNA ensures that most target strands anneal to a primer not to each other.

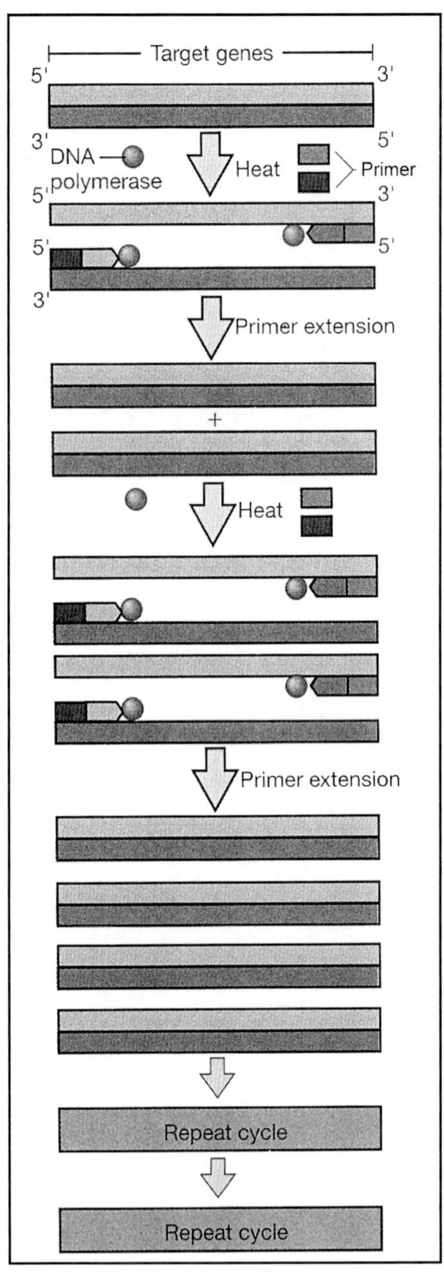


Polymerase Chain Reaction (PCR)

c. DNA polymerase then extends the primers using target strands as template.

d. After appropriate cool, the mixture again heat for denaturation and repeat the above again and again for several cycles.

Thus each PCR cycle involves "Denaturation", "Annealing" and "Primer extension" resulted several fold synthesis of DNA under *in vitro* condition.

*Application of genetic engineering*

Genetic engineering has many commercial and practical applications. The main applications of biotechnology in the area of interest for commercial development are:

i. *Microbial fermentation:* Manipulation of micro-organisms to overproduce the metabolites suchas aminoacids, antibiotics, etc.

ii. *Virus vaccines:* As many of the viral vaccines used viral coat proteins as vaccinating materials, genetic engineering of viral coat protein genes cloned in bacteria, nonpathogenic virus or in plants make the viral vaccines development in safe and convenient.

iii. *Mammalian proteins:* A number of mammalian proteins are greater medicinal and commercial interest are presently allowed to produce by prokaryotic systems in cheaper ways.

iv. *Transgenic crops:* By introducing cloned DNA directly to the plant cells grown tissue culturally and make expressions known as transgenic crops, are of very economically viable technology. The most popular transgenic crops, known as Bt-crops, which harbour the gene *cryIA* coding crystal proteins, a toxin against lepidoptran insects, obtained from bacterium, *Bacillus thuringiensis*. These Bt-transgenic crops are performing better in terms of insect resistance and reducing pesticide application.

v. *Environmental Biotechnology:* Because of the enormous metabolic diversity of bacteria, a large pool exists in bacteria from natural habitats. Many bacteria able to produce metabolites / proteins against environmental pollutants. Such genetic sources are presently used with the help of gene cloning to detoxify the pollutants, wastes, waste water treatment, degradation of recalcitrant chemicals suchas pesticides, toluenes, anilines, hydrocarbons etc.

vi. *Gene therapy:* Controlling the genetic disorders and diseases by using gene cloning / genetic engineering is present advanced medical research known as "Gene therapy".

CHAPTER 16

# Environmental Microbiology

Microorganisms live as single cells in nature in association with other cells in assemblage that we call "Population". Such populations are composed of group of related cells generally derived by successive cell divisions from a single parent cell. The location in an environment where a population generally lives in association with other populations of cells called "Microbial communities". Environmental microbiology deals with the study of these microbial communities in different habitats, naturally occurring on earth.

Natural habitats of microorganisms are exceedingly diverse. Any habitat that is suitable for the growth of microorganisms can also support the growth of other microorganisms. But in addition, there are many habitats where, extreme environment, due to chemical and physical factors, higher organisms cannot grow but microorganisms can grow and flourish. In this chapter, different habitats of microorganisms are discussed briefly.

Because microorganisms are so small, their habitats are also small. The term microenvironment is often used in ecology to describe where a microorganism actually lives and metabolites within its habitat.

## Terrestrial Environments

Terrestrial environment always refers the soil and plant habitats. In the soil on or near the plants, many key processes occur that influence the functioning of the ecosystem. The plant roots penetrate into the soil and their excretions promote the development of rhizosphere microflora.

a. *Soil:* The vast extensive microbial growth takes place on the surface of the soil particles, usually within rhizosphere. Major factors affecting microbial activity in soil is the availability of water, organic carbon status, nutrient status, oxygen, pH and so on. The community structure of microorganisms in soil decides many important elements transformation, which is the key factor for soil fertility in agriculture. They are also source of soil enzymes, which governs the soil biochemical properties.

b. *Deep subsurface microbiology:* The deep soil subsurface terrestrial environment, which can extend for several hundred meters below the soil surface harbor variety of microorganisms, primarily bacteria. The bacteria include anaerobes such as sulphate reducing bacteria, methanogens, homoacetogens are predominant. Compare to surface microorganisms, deep subsurface organisms are low in their metabolic activities. Studying these organisms will be more useful for *in situ* bioremediation of toxic chemicals. Bioremediation refers the use of microorganisms to remove or detoxify toxic or unwanted chemicals in an environment.

## Aquatic Environments

Typical aquatic environments are the oceans, lakes, ponds, rivers, salt marshes, springs etc. Aquatic environments differ considerably in chemical and physical properties and in turn the microbial community also differs. The predominant phototrophic organisms in most aquatic environments are microorganisms; in oxic areas cyanobacteria and algae prevail, an in anoxic areas anoxygenic phototrophic bacteria are predominant. Algae floating or suspended freely in the water is called phytoplanckton, those attached to the bottom or sides are called benthic algae. Because these organisms use energy from light in the initial production of organic matter, they are called as primary producers.

a. *Marine environment:* In the oceans, primary productivity is very low due to limit of nutrients like nitrogen and iron, which limit the phytoplanckton population. A consequence of relatively low primary productivity in the oceans is low heterotrophic microbial activity. This in turn ensures that most ocean waters are oxic, even in greater depth.

b. *Deep sea:* Deep sea refers the zone where visible light is absent (below 300 m depth) in the open ocean water. Organisms that inhabit the deep sea are faced three major environmental factors as low temperature, high pressure and low nutrients level. Most of the microbes of deep sea are psychrophilic and some are extreme psychrophilic. These organisms must also withstand the enormous hydrostatic pressure associated with great depth.

c. *Lakes and rivers:* Significant photosynthetic production of oxygen occur only in the surface layer of lakes and ponds where light is available. Organic matter is not consumed in this layers sinks to the depths and is decomposed by facultative microorganisms using oxygen dissolved in the water. In lakes, once oxygen is consumed, the deep layer become anoxic; strictly anaerobic microorganisms alone grow. In the bottom layer of lakes and ponds, where the oxygen and light is almost absent, a highly diversified anaerobic heterotrophic organisms especially bacterial communities are predominant. In fresh water ecosystems, the biological oxygen demand (BOD) is the important key factor which decides the microbial community structure.

## Extreme Environments and Extremophiles

The world that we are familiar with is full of oxygen, never too cold nor too hot (most of the visible organisms live in a relatively limited range of temperatures from 5°C to 40°C), and we are protected by our atmosphere from most damaging radiation. This familiar world defines what we think of as normal, and the organisms that live at the edges of this world or beyond it are the 'extremophiles.' Interestingly, most extremophiles are microbes and they may live in places that we might think should be uninhabitable. Such places include the very deep oceans which are very cold, very dark with no light to support photosynthesis, and under incredible pressure. Or perhaps they can be found in bubbling cauldrons of acidic waters of Yellowstone National Park. Microbial life began probably about 3 billion years ago, and the first evolutionary steps began in a world without oxygen, with a different atmosphere unable to provide protection from ultraviolet radiation, and with water bodies which were hot and acidic. It was a physiologically challenging world, but one in which the microbes survived and succeeded, and through their success they began to change the world - adding oxygen to the atmosphere and changing the micro-environments into places that are less stressed.

The extremes still exist out there, often still dominated by the microbial communities. The study of these environments and the organisms that live there has considerable value for us. The anoxic world (including the sulfur-dominated thiobios) is arguably the most extensive habitat in the world, at best only the very top layers of sediments and soils are rich in oxygen, the remainder is anoxic. From here, whether rice paddies or mangrove swamps, comes methane one of the most potent greenhouse gases. As we exhaust those parts of familiar biology that are easily knowable, we move increasingly into the less familiar worlds.

An extremophile is an organism that thrives under "extreme" conditions. The term frequently refers to prokaryotes and is sometimes used interchangeably with *Archaea*. In this module, however, you will find that extremophiles come in all shapes and sizes, and that our understanding of the phylogenetic diversity of extreme habitats increases daily. The term extremophile is relatively anthropocentric. We judge habitats based on what would be considered "extreme" for human existence. Many organisms, for example, consider oxygen to be poisonous. While oxygen is a necessity for life as we know it, some organisms flourish in anoxic environments. We call them *extremophiles*.

Most terms used to describe extremophiles are generally straightforward. They are a combination of the suffix *–phile*, meaning "lover of," and a prefix specific to their environment. For example, *acidophiles* are organisms that love (*phile*) acid (*acido*).

*Common extremophiles are,*

1. *Acidophile:* An organism that grows best at acidic (low) pH values.
2. *Alkaliphile:* An organism that grows best at high pH values.
3. *Anaerobe:* An organism that can grow in the absence of oxygen.
4. *Facultative Anaerobe:* An organism that grows in the presence or in the absence of oxygen.
5. *Obligate Anaerobe:* An organism that cannot grow in the presence of oxygen; the presence of oxygen either inhibits growth or kills the organism.
6. *Endolith:* An organism that lives inside rock or in the pores between mineral grains.

7. *Halophile:* An organism requiring high concentrations of salt for growth.
8. *Methanogen:* An organism that produces methane from the reaction of hydrogen and carbon dioxide, member of the *Archaea*.
9. *Oligotroph:* An organism with optimal growth in nutrient limited conditions.
10. *Piezophile (Barophile):* An organism that lives optimally at high hydrostatic pressure.
11. *Psychrophile:* An organism with optimal growth at temperature 15 °C or lower.
12. *Thermophile:* An organism with optimal growth at temperature 40 °C or higher.
13. *Hyperthermophile:* An organism with optimal growth at temperature 80 °C or higher.
14. *Toxitolerant:* An organism able to withstand high levels of damaging elements (e.g., pools of benzene, nuclear waste).
15. *Xerophile:* An organism capable of growth at very low water activity.

The extreme habitats where these extremophiles thrive are:

- *Alkaline:* broadly conceived as natural habitats above pH 9 whether persistently, or with regular frequency or for protracted periods of time.
- *Acidic:* broadly conceived as natural habitats below pH 5 whether persistently, or with regular frequency or for protracted periods of time.
- *Extremely Cold:* broadly conceived habitats periodically or consistently below 5°C either persistently, or with regular frequency or for protracted periods of time. Includes montane sites, polar sites, and deep ocean habitats.

- *Extremely Hot:* broadly conceived habitats periodically or consistently in excess of 40°C either persistently, or with regular frequency or for protracted periods of time. Includes sites with geological thermal influences such as Yellowstone and comparable locations worldwide or deep-sea vents.
- *Hypersaline:* (high salt) environments with salt concentrations greater than that of seawater, that is, >3.5%. Includes salt lakes.
- *Under Pressure:* broadly conceived as habitats under extreme hydrostatic pressure – i.e. aquatic habitats deeper than 2000 meters and enclosed habitats under pressure. Includes habitats in oceans and deep lakes.
- *Radiation:* broadly conceived as habitats exposed to abnormally high radiation or of radiation outside the normal range of light. Includes habitats exposed to high UV and IR radiation.
- *Without Water:* broadly conceived as habitats without free water whether persistently, or with regular frequency or for protracted periods of time. Includes hot and cold desert environments, and some endolithic habitats.
- *Without Oxygen:* broadly conceived as habitats without free oxygen - whether persistently, or with regular frequency, or for protracted periods of time. Includes habitats in deeper sediments.
- *Altered by Humans:* heavy metals, organic compounds; anthropogenically impacted habitats. Includes mine tailings, oil impacted habitats.
- *Astrobiology:* Addresses life beyond the known biosphere – inclusive of life on other heavenly bodies, in space etc. Includes terraforming.

# Glossary

**Abiogenesis :** Abio-means nonliving; genesis means origin.

**Abiotic :** Pertaining to or characterized by the absence of living organisms.

**Acervulus :** A sexual fruiting body or reproductive structure in a fungus.

**Acid curd :** Mill protein coagulated by acid

**Acid dye :** A dye consisting of an acidic organic grouping of atoms (anion), which is the actively staining part, combined with a metal; the dye has affinity for cytoplasm.

**Acid-fast :** Retaining the initial stain and difficult to decolorize with acid alcohol. A property of certain bacteria.

**Actinomycetes :** Gram positive bacteria that are characterized by the formation of branching filaments.

**Active transport :** The energy requiring pumping of ions or other solutes across a cell membrane from a lower to a higher concentration.

**Adaptive enzyme :** An enzyme produced by an organism in response to the presence of the enzyme's substrate or a related substance. Also called induced enzyme.

**Adenine :** A purine component of nucleosides, nucleotides, and nucleic acids.

**Adenosine triphosphate (ATP) :** A compound of one molecule each of adenine and D-ri-bose and three molecules of phosphoric acid; it plays an important role in energy transformations in metabolism.

**Adenoviruses :** A group of icosahedral double-stranded DNA viruses.

**Aerobe :** An organism that requires oxygen for growth and can grow under an air atmosphere (21% oxygen). Compare anaerobe.

**Aerosol :** Atomized particles suspended in air.

**Aflatoxin :** The toxin produced by some strains of the fungus *Aspergillus flavus*; a carcinogen.

**Agar :** A dried polysaccharide extract of red algae (Rhodophyceae) used as a solidifying agent in microbiological media.

**Agglutination :** The clumping of cells.

**Agglutinin :** An antibody capable of causing the clumping or agglutination of bacteria or other cells.

**Akinetes :** Thick-walled single celled nonmotile asexual resting spores formed by the thickening of the parent cell wall; formed by some cyanobacteria.

**Allergy :** A type of antigen-antibody reaction marked by an exaggerated physiological response to a substance in sensitive individuals.

**Allosteric enzymes :** Regulatory enzymes with a binding or catalytic site for the substrate and a different site (the allosteric site) where a modulator acts.

**Amino acid :** An organic compound containing both amino ($-NH_2$) and carboxyl ($-COOH$) groups.

**Ammonification :** The decomposition of organic nitrogen compounds, e.g., proteins, by microorganisms with the release of ammonia.

**Amphitrichous :** Having a single flagellum at each end of a cell.

**Amylase :** An enzyme that hydrolyzes starch.

**Anabolism :** The synthesis of cell constituents from simpler molecules, usually requiring energy. Compare catabolism.

**Anaerobe :** An organism that does not use $O_2$ to obtain energy, cannot grow under an air atmosphere, and for which $O_2$ is toxic. Compare aerobe.

**Anaerobic respiration :** Respiration under anaerobic conditions in which a terminal electron acceptor other than oxygen is involved.

**Anaplasia :** Structural abnormality in a cell or cells.

**Antagonism :** The killing, injury, or inhibition of growth of one species of microorganism by an other when one organism adversely affects the environment of the other.

**Antheridium :** pl. antheridia. A male gametangium.

**Anthramycin :** An antitumor antibiotic.

**Antibiosis :** An antagonistic association between two organisms in which one is adversely affected.

**Antibiotic :** A substance of microbial origin that has antimicrobial activity in very small amounts.

**Antiobody :** Any of a class of substances (proteins) produced by an animal in response to the introduction of an antigen.

**Antigen :** A substance that when introduced into an animal body stimulates the production of specific substance (antibodies) that react or untie with the substance introduced (antigen).

**Antimicrobial agent :** Any chemical or biologic agent that either destroys or inhibits the growth of microorganisms.

**Antiseptic :** Acting against or opposing sepsis, putrefaction, or decay by either preventing or arresting the growth of microorganisms.

**Antiserum :** Blood serum that contains antibodies.

**Antitoxin :** An antibody capable of uniting with and neutralizing a specific toxin.

**Aperture :** The magnitude of the angle subtended by the optical axis and the outermost rays still covered by the objective.

**Aplanospore :** A nonmotile spore; an abortive zoospore.

**Apoenzyme :** The protein moiety (portion) of an enzyme.

**Apothecium :** A sexual fruiting body in a fungus.

**Archaeobacteria / archaea :** A major group of bacteria that includes the methanogens, the red extreme halophiles, and the thermoacidophiles, and which diverged from other bacteria at a very early stage in evolution. Also called archaebacteria.

**Asepsis :** A condition in which harmful microorganisms are absent. Adjective: aseptic

**Aseptic technique :** Precautionary measures taken to prevent contamination.

**Assay :** The qualitative or quantitative determination of the components of a material, such as a drug.

**Assimilation :** The conversion of nutritive material into protoplasm.

**Attenuation :** A weakening; a reduction in virulence.

**Autoclave :** An apparatus using pressurized steam for sterilization.

**Autolysis :** The disintegration of cells by the action of their own enzymes.

**Autotroph :** A microorganism that uses inorganic materials as a source of nutrients; carbon dioxide is the sole source carbon. Compare heterotroph.

**Auxotrophic mutant :** An organism having a growth requirement of specific nutrients not necessary in the parental strain.

**Axenic culture :** A microorganism of a single species, e.g., a bacterium, fungus, alga, or protozoan, growing in a medium free of other living organisms.

**Bacterial filter :** A special type of filter through which bacterial cells cannot pass.

**Bactericide :** An agent that destroys bacteria.

**Bacteriochlorophyll :** A chlorophyll like pigment possessed by anoxygenic photosynthetic bacteria.

**Bacteriophage :** A virus that infects bacteria and causes the lysis of bacterial cells.

**Bacterium, pl. bacteria :** Any of a group of diverse and ubiquitous prokaryotic single-celled microorganisms.

**Basic dye :** A dye consisting of a basic organic grouping of atoms (cation), which is the actively staining part, combined with an acid, usually inorganic; the dye has affinity for nucleic acids.

**Bergey's Manual :** An international reference work, which classifies and describes bacteria.

**Biochemical oxygen demand (BOD) :** A measure of the amount of oxygen consumed in biological processes that break down organic matter in water; a measure of the organic pollutant load.

**Biodegradable :** Capable of being broken down by microorganisms.

**Biogenesis :** The production of living organisms only from other living organisms.

**Bioluminescence :** The emission of light by living organisms.

**Biomass :** The mass of living matter present in a specified area.

**Biosphere :** The zone of the earth that includes the lower atmosphere and the upper layers of soil and water.

**Biota :** The animal, plant, and microbial life characterizing a given region.

**Botulism :** Food poisoning due to the toxin of *Clostridium botulinum*.

**Brownian motion :** A peculiar dancing motion exhibited by finely divided particles and bacteria in suspension, due to bombardment by the molecules of the fluid.

**Budding :** A form of asexual reproduction typical of yeast, in which a new cell is formed as an outgrowth from the parent cell.

**Buffer :** Any substance in a fluid that tends to resist the change in pH when acid or alkali is added.

**Calorie :** A unit of heat; the amount of heat required to raise the temperature of 1g of water by 1°C.

**Capsid :** The protein coat of a virus.

**Capsule :** An envelope or slime layer surrounding the cell wall of certain microorganisms.

**Carotenoid :** A water insoluble pigment, usually yellow, orange or red which consists of along aliphatic polyene chain composed of isoprene units.

**Catabolism :** The dissimilation, or breakdown, of complex organic molecules, releasing energy. A part of the total process of metabolism. Compare anabolism.

**Catalyst :** Any substance that accelerates a chemical reaction but remains unaltered in form and amount.

**Cell :** The microscopic, functionally and structurally basic unit of all living organisms.

**Cell wall :** A rigid external covering of the cytoplasmic membrane.

**Chemoautotroph :** An organism that obtains energy by oxidizing inorganic compounds. Carbon dioxide is the sole source of carbon.

**Chemolithotroph :** An organism that uses in organic compounds as electron donors and relies on chemical compounds for energy.

**Chemotaxis :** The movement of an organism in response to a chemical stimulus.

**Chemotroph :** An organism that uses chemical compounds for energy. Compare phototroph.

**Chlorophyll :** A light trapping green pigment essential as an electron donor in photosynthesis.

**Chloroplast :** A cell plastid (specialized organelle) in plants and algae that contains chlorophyll pigments and functions in photosynthesis.

**Chromosome :** A gene-containing filamentous structure in a cell nucleus; the number of chromosomes per cell nucleus is constant for each species.

**Clone :** A population of cells descended from a single cell.

**Codon :** A sequence of three nucleotide bases (in Mrna) that codes for an amino acid or the initiation or termination of a polypeptide chain.

**Coenzyme :** The non-protein of an enzyme.

**Colony :** A macroscopically visible growth of microorganisms on a solid culture medium.

**Commensalism :** A relationship between members of different species living in proximity (the same cultural environment) in which one organism benefits from the association but the other is not affected.

**Conidium :** An asexual spore that may be one-celled or many-celled and may be of many size and shapes. Also called conidiospore.

**Conjugation :** A mating process characterized by the temporary fusion of the mating partners and the transfer of genes. Conjugation occurs particularly in unicellular organisms.

**Contamination :** The entry of undesirable organisms into some material or object.

**Cyst :** A thick-walled dormant form of an organism which is resistant of desiccation, e.g., the cysts formed by certain bacteria such as *Azotobacter* or by various protozoa.

**Cytochrome :** One of a group of iron porphyrins that serve as reversible oxidation reduction carriers in respiration.

**Cytoplasm :** The living matter of a cell between the cell membrane and the uncleus.

**Cytoplasmic membrane :** A thin layer under the cell wall consisting mainly of phospholipids and proteins; it is responsible for the selective permeability properties of the cell. Also called plasma membrane.

**Dark-field microscopy :** A type of microscopic examination in which the microscopic field is dark and any objects, such as organisms, are brightly illuminated.

**Deoxyribonucleic acid (DNA) :** The carrier or genetic information; a type of nucleic acid occurring in cells, containing phosphoric acid, D-2-deoxyribose, adenine, guanine, cytosine, and thymine.

**Deoxyribose :** A five carbon sugar having one oxygen atom fewer than the parent sugar, ribose; a component of DNA.

**Differential stain :** A procedure using a series of dye solutions or staining reagents to bring out differences in microbial cells.

**Diploid :** Having chromosomes in pairs the members of which are homologous; having twice the haploid number.

**Disinfectant :** An agent that frees from infection by killing the vegetative cells of microorganisms.

**Ecology :** The study of the interrelationships that exist between organisms and their environment.

**Ecosystem :** A functional system which includes the organisms of a natural community together with their environment.

**Endoplasmic reticulum :** An extensive array of internal membranes in a eucaryotic cell.

**Enzyme :** An organic catalyst produced by an organism.

**Epidemiology :** The study of the occurrence and distribution of disease.

**Etiology :** The study of the cause of a disease

**Eubacteria :** One of the two major groups of bacteria (the other being the archaeobacteria). Eubacteria have fundamental features that are considered to be typical of most bacteria.

**Eukaryote :** A cell that possesses a definitive or true nucleus. Compare prokaryote.

**Exponential phase (log phase) :** The period of culture growth when cells divides steadily at a constant rate. Also called logarithmic phase

**Filter bacteriological :** A special type of filter through which bacterial cells cannot pass.

**Fission, binary :** A single nuclear division followed by the division of the cytoplasm to form two daughter cells of equal size.

**Flagellum pl. flagella :** A thin, filamentous appendage on cells, responsible for swimming motility.

**Fluorescence microscopy :** Microbiology in which cells or their components are stained with a fluorescent dye and thus appear as glowing objects against a dark background.

**Fungus, pl. fungi :** A microorganisms that lacks chlorophyll and is usually filamentous; a mold or yeast.

**Gene :** A segment of a chromosome, definable in operational terms as the repository of a unit of genetic information.

**Generation time :** The time interval necessary for a cell to divide.

**Genome :** A complete set of genetic material; i.e., a complete set of genes.

**Glogi apparatus :** A membranous organelle in the endoplasmic reticulum of the cell.

**Gram negative bacteria :** Bacteria that do not retain the crystal violent-iodine complex when subjected to the Gram technique and thus acquire the color of the dye (usually a red dye) that is used to counterstain the cells.

**Guanine :** A purine base, occurring naturally as a fundamental component of nucleic acids.

**Habitat :** The natural environment of an organism.

**Halophile :** A microorganism whole growth is accelerated by or dependent on high salt concentrations.

**Heterotroph :** A microorganism that is unable to use carbon dioxide as its sole source of carbon and requires one or more organic compounds. Compare autotroph.

**Holoenzyme :** A fully active enzyme, containing an apoenyme and a coenzyme.

**Incubation :** In microbiology, the subjecting of cultures of microorganisms to conditions (especially temperatures) favourable to their growth.

**Inhibition :** In microbiology, the prevention of the growth or multiplication of microorganisms.

**Inoculation :** The artificial introduction of microorganisms or other substance into the body or into a culture medium.

**Intercellular :** Between cells.

**Intracellular :** Within a cell.

**Lag phase :** The period of slow, orderly growth when a medium is first inoculated with a culture.

**Lipase :** An enzyme that catalyzes the hydrolysis of fats into glycerol and fatty acids.

**Logarithmic phase :** Commonly called log phase. See exponential phase.

**Lyophilization :** The preservation of biological specimens by rapid freezing and rapid dehydration in a high vacuum.

**Medium, pl. media :** A substance used to provide nutrient for the growth and multiplication of microorganisms.

**Meiosis :** A process occurring during cell division at different points in the life cycles of different organisms, in which the chromosome number is reduced by half, thus compensating for the chromosome doubling effect of fertilization compare mitosis.

**Mesophile :** A bacterium growing best at the moderate temperature range 25 to 40°C.

**Mesosomes :** Membrane invaginations in the form of convoluted tubules and vesicles.

**Messenger RNA :** The intermediary substance that passes information from the DNA in the nuclear region to the ribosomes in the cytoplasm. Abbreviation: mRNA.

**Metabolism :** The system of chemical changes by which the nutritional and functional activities of an organism are maintained.

**Metabolite :** Any chemical participating in metabolism; a nutrient.

**Microbe :** Any microscopic organism; a microorganism. Adjective: microbial.

**Microbiology :** The study of organisms of microscopic size (microorganisms), including their culture, economic importance, pathogenicity, etc.

**Microorganism :** Any organism of microscopic dimensions.

**Mitochondrion :** A cytoplasmic organelle in eukaryotic cells; the site of cell respiration.

**Mitosis :** A form of nuclear division characterized by complex chromosome movement and exact chromosome movement and exact chromosome duplication. Compare meiosis.

**Mold :** A fungus characterized by a filamentous structure.

**Morphology :** The branch of biological science that deals with the study of the structure and form of living organisms.

**Mutation :** A stable change of gene such that the changed condition is inherited by offspring cells.

**Mycelium :** A mass of threadlike filaments, branched or composing a network, that constitutes the vegetative structure of a fungus.

**Mycology :** The study of fungi.

**Nomenclature :** Any system of scientific names, such as those employed in biological classification.

**Nucleic acid :** One of a class of molecules composed of joined nucleotide complexes; the types are deoxyribonucleic acid (DNA) and ribonucleic acid (RNA).

**Nucleolus, pl. nucleoli :** A small body in a cell nucleus.

**Nucleoside :** A pentose sugar linked to a purine or pyrimidine base.

**Nucleotide :** The basic building block of nucleic acids (DNA and RNA); consists of a purine or pyrimidine base, ribose or deoxyribose, and phosphate.

**Nucleus :** The structure in a cell that contains the chromosomes.

**Osmosis :** The passage, due to osmotic pressure, of a fluid through a semipermeable membrane.

**Oxidation :** 1. The process of combining with oxygen. 2. The loss of electrons or hydrogen atoms.

**Pasteurization :** The process of heating a liquid food or beverage to a controlled temperature to enhance the keeping quality and destroy harmful microorganisms.

**Pathogen :** An organism capable of producing disease.

**Periplasmic space :** The space between the cytoplasmic membrane and the outer membrane of Gram-negative bacteria.

**Periplast :** A surface membrane or pellicle of certain algae and bacteria.

**pH :** A symbol for the degree of acidity or alkalinity of a solution; pH = log $(1/[H^+])$, where $[H^+]$ represents the hydrogen ion concentration.

**Phosphorylation :** The addition of a phosphate group to a compound.

**Photolysis :** Light generated break down of water

**Phototroph :** A bacterium capable of utilizing light energy for metabolism.

**Phylogeny :** The evolutionary or ancestral history of organisms.

**Phylum :** A taxon consisting of a group of related classes.

**Physiology :** The study of the life processes of living things.

**Pilus :** Any filamentous appendage other than flagella on certain Gram-negative bacteria.

**Prokaryote :** A type of cell in which the nuclear substance is not enclosed within a membrane; e.g., a bacterium or cyanobacterium. Compare Eukaryote.

**Protoplasm :** The living substance of a cell. The term usually refers to the substance enclosed by the cytoplasmic membrane.

**Psychrotroph :** An organism which is able to grow at 0°C but which grows best at a temperature above 20°C; also termed facultative psychrophile.

**Reduction :** A chemical process involving the removal of oxygen, the addition of hydrogen atoms, or the gain of electrons.

**Respiration :** An energy-yielding process in which electrons from an oxidizable substrate are transferred via a series of oxidation-reduction reactions to an exogenous terminal electron acceptor.

**Rhizosphere :** The soil region subject to the influence of plant roots and characterized by a zone of increased microbiological activity.

**Ribonucleic acid (RNA) :** A nucleic acid occurring in the cytoplasm and the nucleolus, containing phosphoric acid, D-ribose, adenine, guanine, cytosine and uracil.

**Ribosomal RNA (rRNA) :** The RNA of the ribosomes constituting about 90 per cent of the total cellular RNA.

**Ribosome :** A cytoplasmic structural unit, made up of RNA and protein that is the site of protein synthesis.

**Saprophyte :** An organism living on dead organic matter.

**Species :** The basic taxonomic group; in bacteriology, a species consists of a type strain together with all the other strains that are considered sufficiently similar to the type strain to warrant inclusion in the species.

**Spore :** A resistant body formed by certain microorganisms.

**Sterilization :** The process of making sterile; the killing of all forms of life.

**Substrate :** The substance acted upon by an enzyme.

**Symbiosis :** The living together of two or more organisms; microbial association

**Synergism :** The ability of two or more organisms to bring about changes (usually chemical) that neither can accomplish alone.

**Taxonomy :** The classification (arrangement), nomenclature (naming), and identification of organisms.

**Thermophile :** An organism that grows best at temperatures above 45°C.

**Transcription :** The process in which a complementary single stranded mRNA is synthesized from one of the DNA strands.

**Transduction :** The transfer of genetic material from one bacterium to another through the agency of a virus.

**Transfer RNA (tRNA) :** A specific RNA for each amino acid that becomes esterified to the terminal adenosine. Each of the 60 or so tRNAs has a specific trinucleotide sequence that interacts with a complementary sequence in mRNA. Also called soluble RNA (sRNA).

**Transformation :** The phenomenon by which certain bacteria incorporate DNA from related strains into their genetic makeup.

**Tyndallization :** A process of fractional sterilization with flowing steam.

**Ultraviolet rays :** Radiations from about 3900 to about 2000 Å.

**Vaccine :** A preparation of killed or attenuated microorganisms, or their components, or their products that is used to induce active immunity against a disease.

**Virion :** The complete mature virus particle.

**Virucide :** An agent that kills viruses.

**Virus :** An obligate intracellular parasitic microorganism that is smaller than bacteria. More viruses can pass through filters that retain bacteria.

**Yeast :** A kind of fungus that is unicellular and lacks typical mycelia.

**Zygospore :** A kind of spore resulting from the fusion of two similar gametes in some fungi.

**Zygote :** An organism produced by the union of two gametes.

# Colour Versions of Original Plates in Text

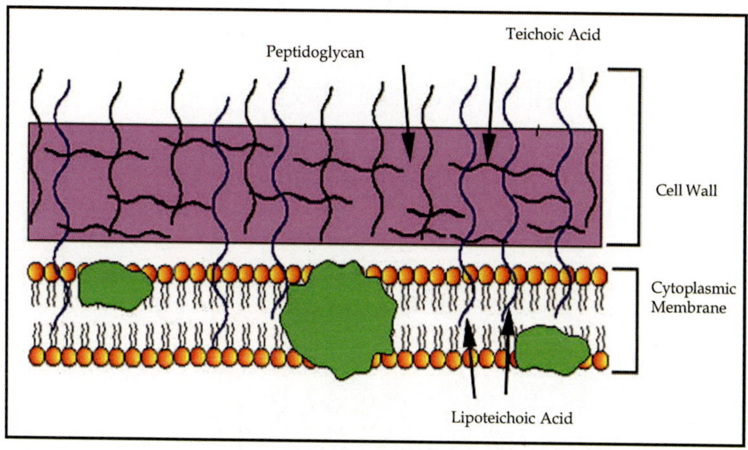


*[For the original version of this figure see page 66]*

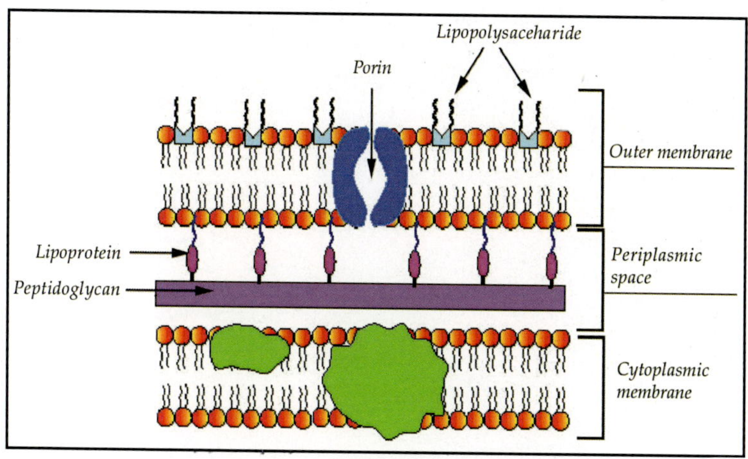


Structural arrangement of outer layer

*[For the original version of this figure see page 68]*

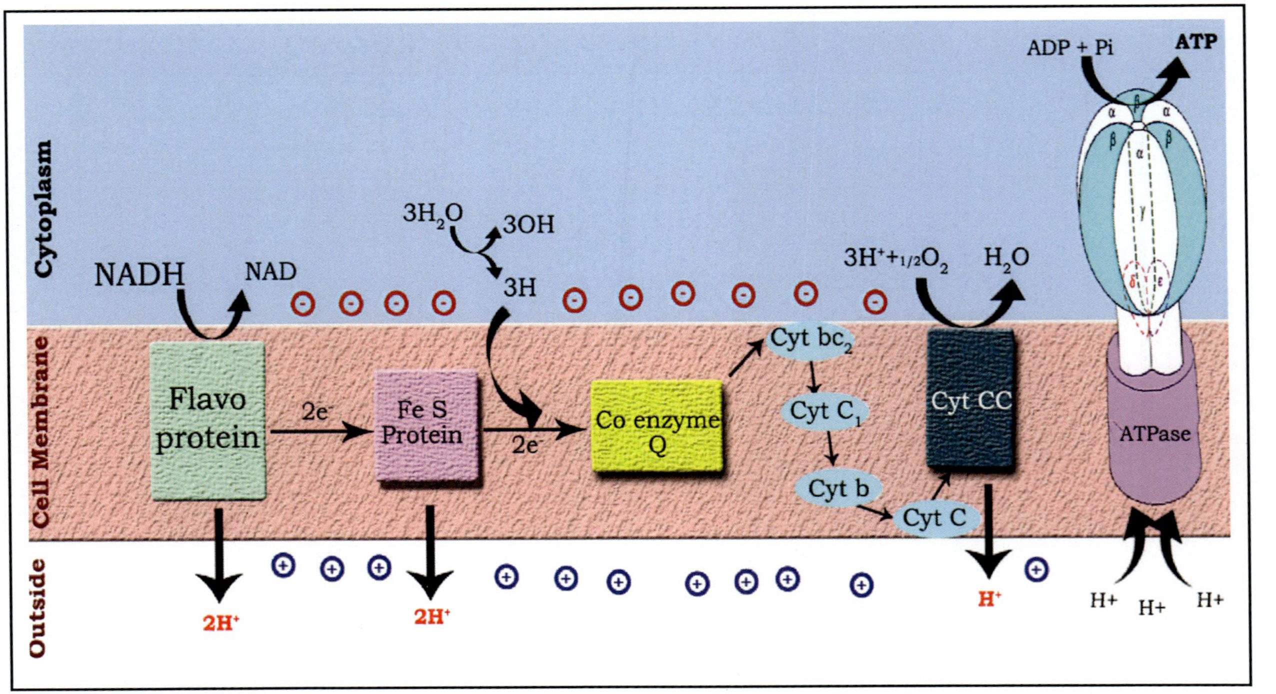


*[For the original version of this figure see page 195]*

# Index

**C**



**D**



**E**